Algorithms for Intelligent Systems

This book series publishes research on the analysis and development of algorithms for intelligent systems with their applications to various real world problems. It covers research related to autonomous agents, multi-agent systems, behavioral modeling, reinforcement learning, game theory, mechanism design, machine learning, meta-heuristic search, optimization, planning and scheduling, artificial neural networks, evolutionary computation, swarm intelligence and other algorithms for intelligent systems.

The book series includes recent advancements, modification and applications of the artificial neural networks, evolutionary computation, swarm intelligence, artificial immune systems, fuzzy system, autonomous and multi agent systems, machine learning and other intelligent systems related areas. The material will be beneficial for the graduate students, post-graduate students as well as the researchers who want a broader view of advances in algorithms for intelligent systems. The contents will also be useful to the researchers from other fields who have no knowledge of the power of intelligent systems, e.g. the researchers in the field of bioinformatics, biochemists, mechanical and chemical engineers, economists, musicians and medical practitioners.

The series publishes monographs, edited volumes, advanced textbooks and selected proceedings.

More information about this series at http://www.springer.com/series/16171

Jitendra Kumar Rout · Minakhi Rout ·
Himansu Das
Editors

Machine Learning for Intelligent Decision Science

Springer

Editors
Jitendra Kumar Rout
School of Computer Engineering
Kalinga Institute of Industrial Technology
Deemed to be University
Bhubaneswar, Odisha, India

Minakhi Rout
School of Computer Engineering
Kalinga Institute of Industrial Technology
Deemed to be University
Bhubaneswar, Odisha, India

Himansu Das
School of Computer Engineering
Kalinga Institute of Industrial Technology
Deemed to be University
Bhubaneswar, Odisha, India

ISSN 2524-7565 ISSN 2524-7573 (electronic)
Algorithms for Intelligent Systems
ISBN 978-981-15-3688-5 ISBN 978-981-15-3689-2 (eBook)
https://doi.org/10.1007/978-981-15-3689-2

This Springer imprint is published by the registered company Springer Nature Singapore Pte Ltd.
The registered company address is: 152 Beach Road, #21-01/04 Gateway East, Singapore 189721, Singapore

Preface

Decision science is the process of selecting logically a best choice from the available options to make an appropriate decision. One must need to weigh the pros and cons of each option as well as all the alternatives to make an appropriate decision. Decision science analyzes a large amount of data for a particular domain which is a very tedious task for handling manually. For effective decision-making, a technique must be able to forecast the outcome of each option as well as to determine which option is the best for a particular situation. Machine learning algorithms can efficiently handle a large amount of data to build mathematical models in order to make predictions or decisions without being explicitly programmed to perform the task.

Machine Learning (ML) is the study of algorithms and mathematical models that computer systems use to progressively improve their performance on a specific task. Machine learning-based decision-making model develops new, intelligent, hybrid, and adaptive methods and tools for solving complex learning and decision-making problems under conditions of uncertainty. Machine learning is widely used in various domains to perform various tasks effectively to analyze and process huge amount of data for predictive analytics, recommendations, classification, clustering, feature learning, dimensionality reduction, pattern recognition, and information retrieval in less amount of time with greater accuracy.

Decision science in bioinformatics is to develop computational methods to analyze large collections of biological data to discover sequence alignment, gene finding, genome assembly, protein structure alignment and its prediction, the prediction of gene expressions and protein–protein interactions, and the modeling of evolution. In financial domain, decision can be for risk assessment, trend analysis, portfolio management, interest rate prediction, etc. In recommendation systems, it is to analyze user profiles to generate personalized recommendations where such profiles are often too coarse to capture the current user's state of mind/desire. For natural language processing, decision-making is to program computers to process and analyze large amounts of natural language data like speech and text. Similarly, in digital image processing decision is to carry out automatic processing, manipulation, and interpretation of such visual information, and it plays an increasingly important role in many aspects of our daily life, as well as in a wide variety of disciplines and fields in science and technology, with applications such as television, photography, robotics, remote sensing, medical diagnosis and industrial inspection, and cloud analysis.

The objective of this edited book is to provide all aspects of computational intelligence methods to develop efficient, adaptive, and intelligent models to handle the challenges related to decision-making in various aspects which help the researchers to take this to the next level. It also provides a platform for data scientists, practitioners, and educators to share the most recent trends, practical challenges, and advances in the field of machine learning and intelligent decision science. By looking at its popularity and application in interdisciplinary research fields, this book focuses on the advances and applications of machine learning and its usefulness in decision-making process in various aspects.

In Chap. 1, roy et al. addresses various types of geo-environmental problems in the fringing area of Chhotanagpur Plateau in India, and gully erosion is one of them. In Chap. 2, authors focus on a new deep CNN (11-layer) model for automatically classifying ECG heartbeats into five different groups according to the ANSI-AAMI standard (1998) without using feature extraction and selection techniques. Chapter 3 reviews and presents various machine learning and deep learning algorithms for disease identification. Chapter 4 presents an interactive PSO-GA algorithm that performs parallel processing of PSO and GA using multi-threading and shared memory for information exchange to enhance convergence time and global exploration. In Chap. 5, author presents the root cause analysis model for effective decision-making. This model consists of multiple models, namely, aspect categorization ontology for aspect extraction, prediction-based word embedding model, variegated ensemble-based weighted voting model for prediction. It is used to reduce the computational complexity and error, and ontology reinforcement for frequent updates in the ontology system. Chapter 6 presents the details of the nuances of SMO specifically the phases involved, namely, the leader phase, learning phase, and decision phase. It also introduces the basic mathematical jargon and fundamentals that are required to model an SMO algorithm for finding the optimal solution to any in-hand problems. Various variants of SMO are also covered in this chapter with a detailed overview of the pros and cons of each of the variants focusing on the research gaps. In Chap. 7, authors address the need for and usefulness of MAS by giving the reader an insight into the agents' characteristics, its interaction with the environments, various performance measures, and different types of MAS. Chapter 8 presents the development of robust computer-assisted malaria diagnosis in light microscopic blood images.

Topics presented in each chapter of this book are unique to this book and are based on unpublished work of contributed authors. In editing this book, we attempted to bring into the discussion all the new trends and experiments that have made on machine learning using intelligent decision-making process. We believe this book is ready to serve as a reference for a larger audience such as system architects, practitioners, developers, and researchers.

Bhubaneswar, Odisha, India Himansu Das

Contents

About the Editors

Jitendra Kumar Rout is an Assistant Professor at the School of Computer Engineering, KIIT Deemed to be University, Bhubaneswar, India. He completed his Masters and Ph.D. at the National Institute of Technology, Rourkela, India, in 2013 and 2017 respectively, and was a lecturer at various engineering colleges, such as GITA and TITE Bhubaneswar. He is a life member of Odisha IT Society (OITS) and has been actively involved in conferences like ICIT (one of the oldest conferences in Odisha). He is also a life member of IEI, and a member of IEEE, ACM, IAENG, and UACEE. His main research interests include data analytics, machine learning, NLP, privacy in social networks and big data, and he has published his work with IEEE and Springer.

Minakhi Rout is currently an Assistant Professor at the School of Computer Engineering, KIIT Deemed to be University. She received her M. Tech and Ph.D. degrees in Computer Science and Engineering from Siksha 'O' Anusandhan University, Odisha, India, in 2009 and 2015, respectively. She has more than 13 years of teaching and research experience at various respected institutes, and her interests include computational finance, data mining and machine learning. She has published more than 25 research papers in various respected journals and at international conferences. She is editor for the Turkish Journal of Forecasting.

Himansu Das is an Assistant Professor at the School of Computer Engineering, Kalinga Institute of Industrial Technology (KIIT) Deemed to be University, Bhubaneswar, India. He holds a B. Tech degree from the Institute of Technical Education and Research, India and an M. Tech degree in Computer Science and Engineering from the National Institute of Science and Technology, India. He has published several research papers in various international journals and at conferences. He has also edited several books for leading international publishers like IGI Global, Springer and Elsevier. He serves as a member of the editorial, review or advisory boards of various journals and conferences. Further, he has served as organizing chair, publicity chair and member of the technical program committees of several national and international conferences. He is also associated with various educational and research societies like IET, IACSIT, ISTE, UACEE, CSI, IAENG, and ISCA. He has more than 10 years of teaching and research experience, and his interests include data mining, soft computing and machine learning.

Chapter 1
Development of Different Machine Learning Ensemble Classifier for Gully Erosion Susceptibility in Gandheswari Watershed of West Bengal, India

Paramita Roy, Rabin Chakrabortty, Indrajit Chowdhuri, Sadhan Malik, Biswajit Das, and Subodh Chandra Pal

Abstract In various types of geo-environmental problems in the fringing area of Chhotanagpur plateau in India, gully erosion is one of the vulnerable issue. In our current research, using Multi-layer perception approach (MLPC) and its ensembles (MLPC-Bagging, MLPC-Dagging and MLPC-Decorate) models, we have identified potentiality zone of gully erosion in Gandheswari watershed. Considering 20 geo-environmental factors, namely; rainfall, slope, slope aspect, elevation, drainage density, Land use and land cover (LULC), Normalized difference vegetation index (NDVI), geology, geomorphology, soil texture, soil moisture, distance from road, distance from river, plan curvature, profile curvature, topographical wetness index, stream power index, terrain ruggedness index, soil erodibility and distance from lineament, the susceptible areas are indentified. The five susceptible zones are identified with the help of the MLPC computational approaches and different ensemble classifier. All the models are predicted of fitted with good manner but the MLPC-Decorate is comparatively better than other models. The Area under curve (AUC) values of Receiver operating characteristic (ROC) curve for the MLPC-Decorate model in the training and validation database are 0.924 and 0.906 respectively. This model can be use in any type of environmental modelling in sub-tropical region.

P. Roy · R. Chakrabortty · I. Chowdhuri · S. Malik · B. Das · S. C. Pal (✉)
Department of Geography, The University of Burdwan, Barddhaman, West Bengal, India
e-mail: geo.subodh@gmail.com

P. Roy
e-mail: paramitaroy95@gmail.com

R. Chakrabortty
e-mail: rabingeo8@gmail.com

I. Chowdhuri
e-mail: indrajitchowdhuri@gmail.com

S. Malik
e-mail: sadhanmalik06@gmail.com

B. Das
e-mail: biswajitdas1119@gmail.com

J. K. Rout et al. (eds.), *Machine Learning for Intelligent Decision Science*,
Algorithms for Intelligent Systems, https://doi.org/10.1007/978-981-15-3689-2_1

Keywords Gully erosion · MPLC-Bagging · MPLC-Dagging · MPLC-Decorate · Environmental modelling

1 Introduction

Gully erosion is one of the significant erosional factors of earth surface through water but not dominant. Such type of erosion removes the soils of unprotected land and washes it away through inferior drainage lines and in less vegetated areas. In all naturally occurring process which affects landforms, soil loss or erosion is one of them. Soil erosion is defined as the top-soils of land by natural phenomena like a glacier, wind, water and tidal effects. Soil erosion is a combination of three processes like soil detachment, movement of soil and deposition. Soil compaction, low organic matter, loss of soil structure, poor internal drainage and other soil degradation condition are components which can increase the soil erosion process. From 1980s geographers and researchers are giving more emphasis to predict gully erosion in eastern India river basin. Soil erosion may happen due to several methods such as agriculture, deforestation and other natural processes. The most vulnerable are human activities such as massive number of built-up profiles an increasing population capture the river catchment, and inversely its bad effect goes to on same human or population because gully erosion degraded soil fertility, ecological productivity and destroyed the ecological system [1]. Such type of erosion is a natural process mainly caused by rainfall and another influencing factor which are name as topography, vegetation, lineament and climate in other hands these are the parameter for rainfall happening. If we look at the global distribution of gully erosion, then Asia gets a prominent position because of its monsoon climatic condition [2]. Annually, 35 million hectares soil is removed each year. The Eastern part of India where flash flood is prevalent and happened at very rapid rates, threats the ecosystem by the natural process [3]. Soil erosion is directly interconnected to flash flood due to lack of vegetation. The flash flood is frequent in Gandheswari river basin due to the high amount of sedimentation so, sand splay shows a positive relationship with erosion. Sand splay area's soil is not compacted, and sand is spread over the land and where rainfall creates first sheet flow, and in next stage, water flows as rill erosion over the sand splay area. And this area is under undulation where steepness of slope variation is common to river basin; suddenly changes of the slope which establishes surface run off over the land and removed the surface soil structure which is known as gully erosion. The monsoonal climatic condition, specific geomorphology and other suitable factors are responsible for such type of soil erosion. Long-term dry weather in summer time and short wet period with high intensity of rainfall create surface runoff over than infiltration to the downward. Mekonnen et al. [4] has stated that the natural soil erosion process, gully erosion will be destructive land degradation process if we do not manage or predict this type of erosional area, properly. We have studied lots of literature where many models are used to predict possible erosional area like Naïve Bios Decision Trees (NBDT) [5], Artificial neural network [6] Analysis

Decision Trees (ADT) [7], Frequency Ratio (FR) [8], Analytical Hierarchy Process (AHP) [9–11] Revised Universal Soil Loss Equation (RUSLE), Flexible Discriminant Analysis (FDA), Multivariate Addictive Regression Splines (MARS) and SVM, considering different hydrological and environmental factors with the assistance of remote sensing and GIS. Above the mention machine learning (ML) methods like Particle swarm optimization (PSO), supervised classification based prediction model and Convolutional neural network has been applied by the various researchers in different discipline [12–17]. As well as in the developed country, developing country is also using different RS-GIS techno-centric approach through specific model to delineate gully susceptible area [18].

In sub-tropical area, Gandheswari river basin is situated at Chhota Nagpur plateau where gully erosion is very common in monsoon time and causes soil degradation due to distributed land, agricultural practices, deforestation and human settlement. In this area, soils play a vital role to product food grains. A frequent agricultural practice, uses of chemical fertilizer destroys soil structure and deforestation creates poor drainage system which promotes rill erosion to gully erosion. If it is under monsoonal climatic areas dry summer and dry winter are common characteristics and temperature of winter indicate 10–15 °C and in summer is 27–32 °C, range of temperature is 7–10 °C. The amount of annual precipitation is 111.76 mm. Sometimes total climatic characteristics are influenced by El-NINO. This factor causes delay of coming monsoon and increasing temperature, high intensity of rainfall with short duration are appropriate factors to happen gully erosion in this area.

Akgün and Türk [19] has stated that statistical models of machine learning methods primarily establish gully susceptible maps or soil erosional map which is similar to flood susceptible methods [20]. And these physical models are most important to prepare more reliable susceptible maps of gully erosion with space time evolution. Gully susceptible areas will be considered under one umbrella that is MLPC (multi-layer perception approach and its ensembles are MLPC-Bagging, MLPC-Dagging and MLPC-Decorate. Training dataset will be validated through Kappa and RMSE (Root mean square error). We have viewed the literature of [21] where he has stated gully erosion is the behaviour on the land surface which is considered through Bernoulli probability distribution. With the help of RS-GIS environment, we can easily compare the susceptible areas with real gully erosional area and estimate the intensity and magnitude of that susceptible area in coming years. The modern technique RS-GIS and field observation can implement the entire result of susceptible areas and its objectives with integrated process. We have used DEM with 12.5 m resolution for preparing the environmental factors to consider as training dataset. Major factors like NDVI, LULC had been extracted from Sentinel 2A images with 10 m resolution. To estimate the reality a field study is dine for collecting primary data.

This current research will identify such susceptible area using MLPC (Multilayer Perception Approach) and its ensembles methods (Bagging, Dagging and Decorate) considering 20 factors name as slope, plan curvature, elevation, distance from river, drainage density, lithology, geomorphology, rainfall, TWI, SPI, distance to road, LULC, NDVI, soil texture, soil moisture, aspect, profile curvature, terrain ruggedness index, soil erodibility and distance from the lineament. We have aim to prepare this susceptible maps for managing the soils and give awareness to create sustainability for this environment [22]. It is the trend where geographers, engineer, regional planners and farmers are trying to protect element of environment such as water and soil. Soil is the most important biological factor where people live, product food for live, but humans' rapid interaction disbalances the earth sustainability [23].

2 Study Area

Dwarekeswar River locally known as Dhalkisor is a major in the western part of Indian state of West Bengal [24]. It is one of the 26th river sub-basins of this state and is under the Ganga-Bhagirathi river system. This river originated from Tilaboni hill in Purulia district near Chatna. Gandheswari is a prominent tributary of upper Dwarkswar river basin. Rising from the district of Bankura, Ganheswari River is confluence with Dwarekeswar River near Bankura town [25]. This river basin is located between longitudes 86°53′11″ to 87°08′00″E and latitudes 23°13′15″ to 23°31′25″N and occupies a total area of 392.68 sq.km (Fig. 1). The master slope of this area trends towards the south-east direction [26]. Gandheswari river basin area is located in the district of Bankura, right bank of Damodar River bounded this area from north to north-eastern direction and itss another side is covered by the left bank of Dwarakeswar river from south to south-western part. EIA-EMP report [27], irrigation and water Directorate, Govt. of West Bengal has reported that dam on Dwarakeswar, Gandheswari river and water barrage are located in Bankura, West Bengal. The physiographic condition of this district is undulating and tiny rivulet is very common in here. Western part of Bankura district is covered by lateritic soils. The maximum temperature of summer and minimum temperature of winter are 42 °C and 6 °C respectively. The annual average rainfall varies from 105.5 cm to 107.03 cm and 81% of total rainfall received during the monsoon season. This area is under tropical savanna climate. So, naturally except during rainy season the river basin is become dry in another season. Others soils type such as red soil and brown soil are found here. Colluvial soil and skeletal soil are reached by the amount of coarse and gravel.

3 Database and Methodology

3.1 Used Dataset

This research includes different types of dataset. This are ALOS-PALSAR DEM with 12.5 m spatial resolution, Seninel 2A images with 10 m resolution, Topographical map at 1:50000 scale and Geological map at 1:1000000 scale.

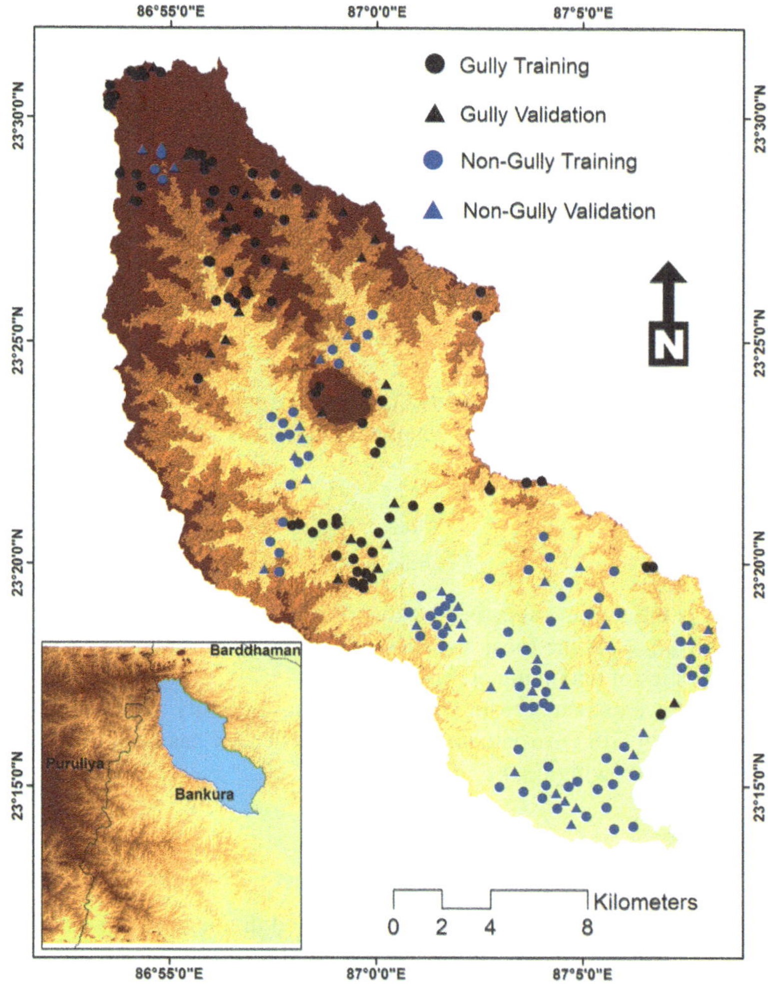

Fig. 1 Location of the study area

3.2 Orientation of the Data

Various topographical, hydrological, soil characteristic, geological and environmental conditioning parameters have developed gully erosion [28]. Based on local condition and several literature review name as Conoscenti et al. [29] these factors were selected, because analysis of all gully conditioning factors are not able to do. Before described 20 parameters were used and calculated. And also have said in the above that elevation, slope, plan curvature, TWI, SPI from ALOS-PALSAR DEM with 12.5 m resolution.

The topographical wetness index is determined with considering the following Eq. 1.

$$\mathrm{TWI} = \mathrm{Ln}\left(\frac{\alpha}{\tan\beta}\right) \tag{1}$$

where α is the cumulative upslope area draining through a point, β is the slope raster and Ln is the function of ArcGIS environment.

The stream power index is determined with considering the following Eq. 2.

$$\mathrm{SPI} = \mathrm{As} \times \tan\beta \tag{2}$$

where, As is the specific catchment area (m2m − 1) and β is the slope raster.

The terrain ruggedness index or roughness index have been estimated with the assist of the subsequent Eq. 3

$$\mathrm{TRI} = \frac{\mathrm{FS_{mean}} - \mathrm{FS_{minimum}}}{\mathrm{FS_{maximum}} - \mathrm{FS_{Minimum}}} \tag{3}$$

where, TRI is the terrain ruggedness index, $\mathrm{FS_{mean}}$ is the focal statistics of mean elevation, $\mathrm{FS_{maximum}}$ is the focal statistics of maximum elevation and $\mathrm{FS_{Minimum}}$ is the focal statistics of maximum elevation.

Drainages have extracted from DEM and drainage density and distance to river have calculated in GIS environment. The drainage density has been estimated with considering the following Eq. 4.

$$D_d = \frac{L_u}{w_a} \tag{4}$$

where, L_u is considered as entire length of selected drainage in km and w_a is the area of the watershed which is estimated in GIS platform. The direct tools of GIS are considered to prepare raster map of drainage density and distance to road. With the help of Google earth images and topographic maps, roads have been extracted. For making the rainfall raster, the primary observation regarding the amount of precipitation forms different rain gauge station has been done during the time of field visit. Sentinel 2A images provides LULC and NDVI for environment factors. Sentinel 2A is optical imaging satellite sensor has been monitored by European Space Agency's Copernicus Programme, which provide high-resolution multispectral images [30].

The NDVI has been determined with allowing the following Eq. 5 [31, 32]:

$$\mathrm{NDVI} = \frac{\mathrm{NIR} - \mathrm{RED}}{\mathrm{NIR} + \mathrm{RED}} \tag{5}$$

Lineament density and lithology of the study area is determined from geological map at 1:1000000 scale. The samples are collected from the field visit and the physical and chemical characteristic are considered for estimating the soil texture and soil erodibility. The soil erodibility factor is determined with allowing the following Eq. 6 [33];

$$K = 0.0137 \times \left(0.2 + 0.3 \times e^{\left[-0.0256 \times \text{San} \times \left(\frac{1 - Sil}{100}\right)\right]}\right) \times \left(\frac{\text{Sil}}{\text{Cla} + \text{Sil}}\right)^{0.3} \times \left[1 - \frac{0.25 \times \text{TOC}}{\text{TOC} + e^{(3.72 - 2.95 \times \text{TOC})}}\right] \times \left[1 - \frac{0.7 \times \text{SN}_1}{\text{SN}_1 + e^{(22.9 \times \text{SN}_1 - 5.51)}}\right] \quad (6)$$

where, K is the soil erodibility which is obtained with considering the soil physical and chemical properties. San is the percentage of sand content, Sil is the percentage of silt content, Cla is the percentage of clay content, TOC is the percentage of soil total organic carbon content and SN_1 is the $1 - \text{San}/100$. The information regarding the soil moisture are collected from the direct primary observation and incorporated in this study.

4 Materials and Methodology

4.1 Geo-Environmental Factors

Determination of geo-environmental parameters for susceptible maps of gully erosional areas is first and important steps [34]. Combined, the geo-environmental data and primary inventory data can prepare a susceptible map of any areas. All factors are conceptually classified in five categories.

4.1.1 Topographical Factors

Topographic factors are geomorphic in character which has great important on soil erosion as well as gully erosion. These factors are collected from topographical sheet at 1:50000 and as well as from ALOS-PALSAR DEM. To show morphometric environmental setting slope, elevation, aspects, plan curvature, profile curvature and TRI compared to topographical aspect (Fig. 2).

4.1.2 Hydrological Factors

Hydrological aspect is named as distance to drainage, rainfall, drainage density, SPI and TWI [35, 36]. The station-wise primary rainfall data have been incorporated in this study for estimating the rainfall raster (Fig. 3).

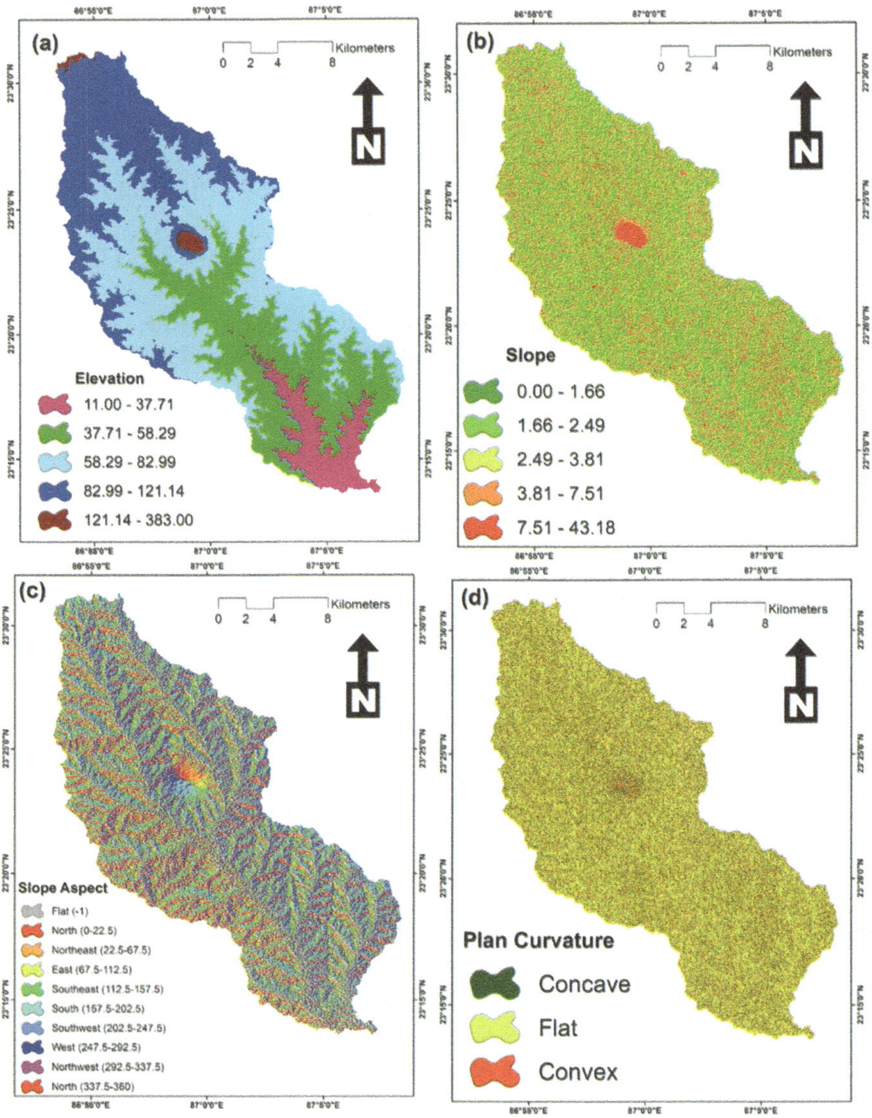

Fig. 2 Topographical factors: Elevation (**a**), slope (**b**), slope aspect (**c**), plan curvature (**d**), profile curvature (**e**) and terrain ruggedness index (**f**)

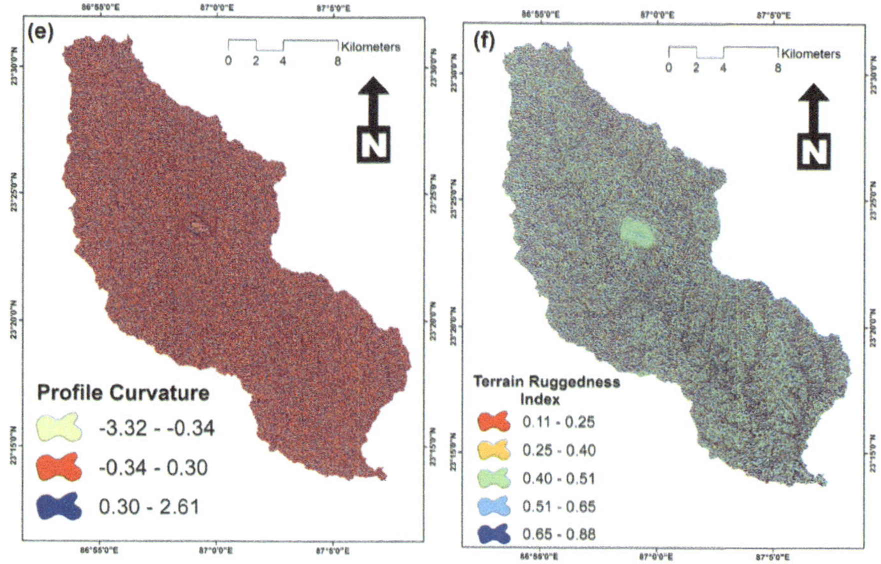

Fig. 2 (continued)

4.1.3 Soil Characteristics

We also counted organic matter, soil texture, soil moisture as soil properties. Different physical and chemical properties are considered to estimate the textural classes and the soil erodibility factor (Fig. 4).

4.1.4 Environmental Factors

LULC, NDVI, and distance to road are considered as environmental factors [37, 38]. Road networking maps are created from topographical maps. Five LULC units are found in this area, these are vegetation, agriculture, built-up area, shrub-land and water body. The distances from road have a five layers and an interval with 1000 m (Fig. 5).

4.1.5 Geological Parameters

To describe the lithology and lineament density of this study area geological survey of India provides the information (Fig. 6).

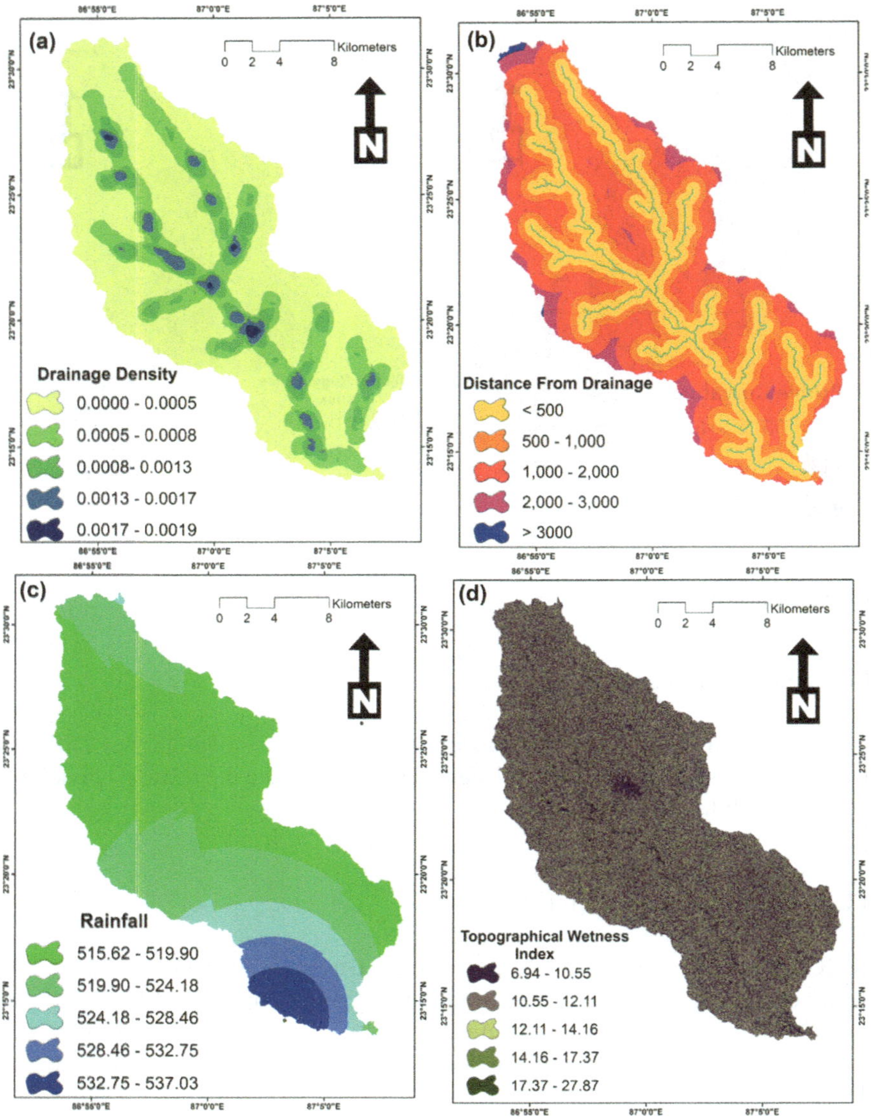

Fig. 3 Hydrological factors: Drainage density (**a**), distance from drainage (**b**), rainfall (**c**), topographical wetness index (**d**), profile curvature (**e**) and stream power index (**f**)

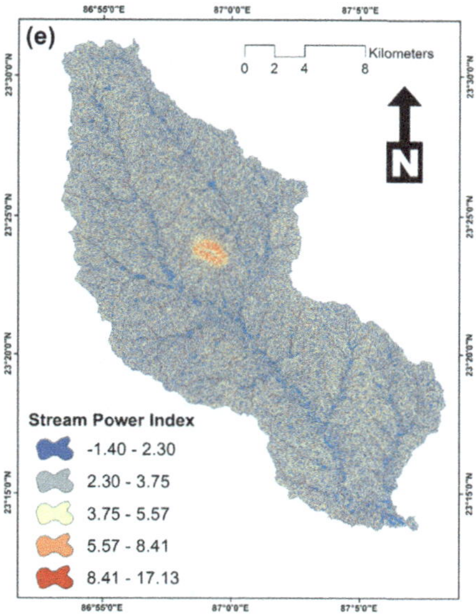

Fig. 3 (continued)

4.2 Gully Erosion Inventory Map

Any inventory maps show the location. Here means we have prepared Gandheswari watershed inventory map where gullies are shown in the map and this map also gives us information about the future trend of gully erosion (Fig. 1). And also shows susceptible areas and degraded areas due to gulling. We have done our empirical observation with Global positing system (GPS) and also used Google earth imagery. Field survey was done using Google earth images and global positioning system also registered the coordination of each gully erosion polygon. Primary data and secondary data were considered through colinearity assessment. Validation data will be considered through Cohen kappas model. Eighty-six gullies and non-gullies were observed in Gandheswari watershed where 70% is considered as model building and 30% is considered as for validation process. We had build gully erosional model by MLPC and Bagging, Dagging, Decorate under one umbrella. By this process we can easily balance primary data and secondary data for susceptible map of gully and it is the same process of other natural hazard susceptible mapping.

4.3 Description of Methodology

Flowchart has shown the methodology where different types of model play an important role. And this methodology has included four parts as data preparation, Multi-colinearity assessment using Variance inflation factors (VIF) and Tolerance

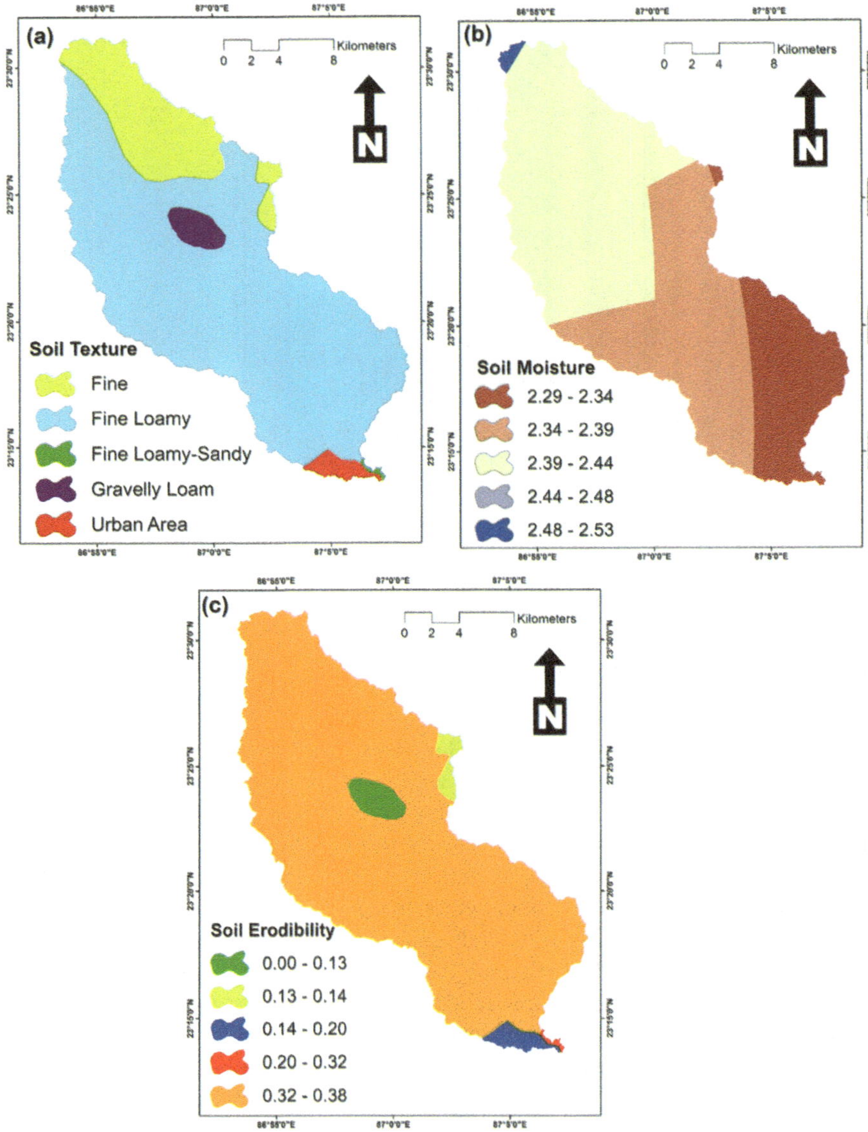

Fig. 4 Hydrological factors: Soil texture (**a**), soil moisture (**b**) and soil erodibility (**c**)

(TOL) and Gully erosion susceptibility modelling by MLPC and its ensembles and Validation process through ROC and Cohen Kappa's model.

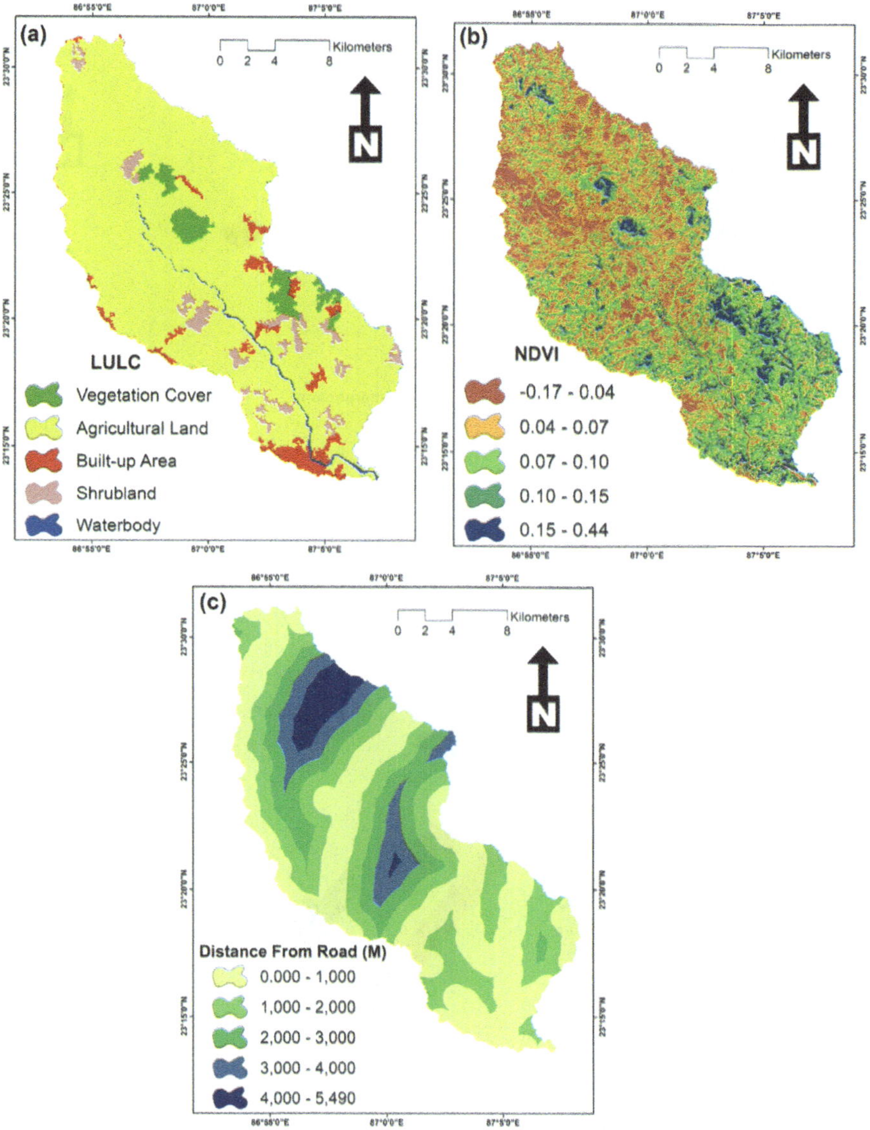

Fig. 5 Hydrological factors: LULC (**a**), NDVI (**b**) and distance from road (**c**)

4.3.1 Flow Chart of Dataset

Different methods and its relevant database have been selected purposively for estimating the gully erosion susceptibility assessment in Gandheswari watershed of West Bengal, India. The detailed methodology regarding this work is given in Fig. 7.

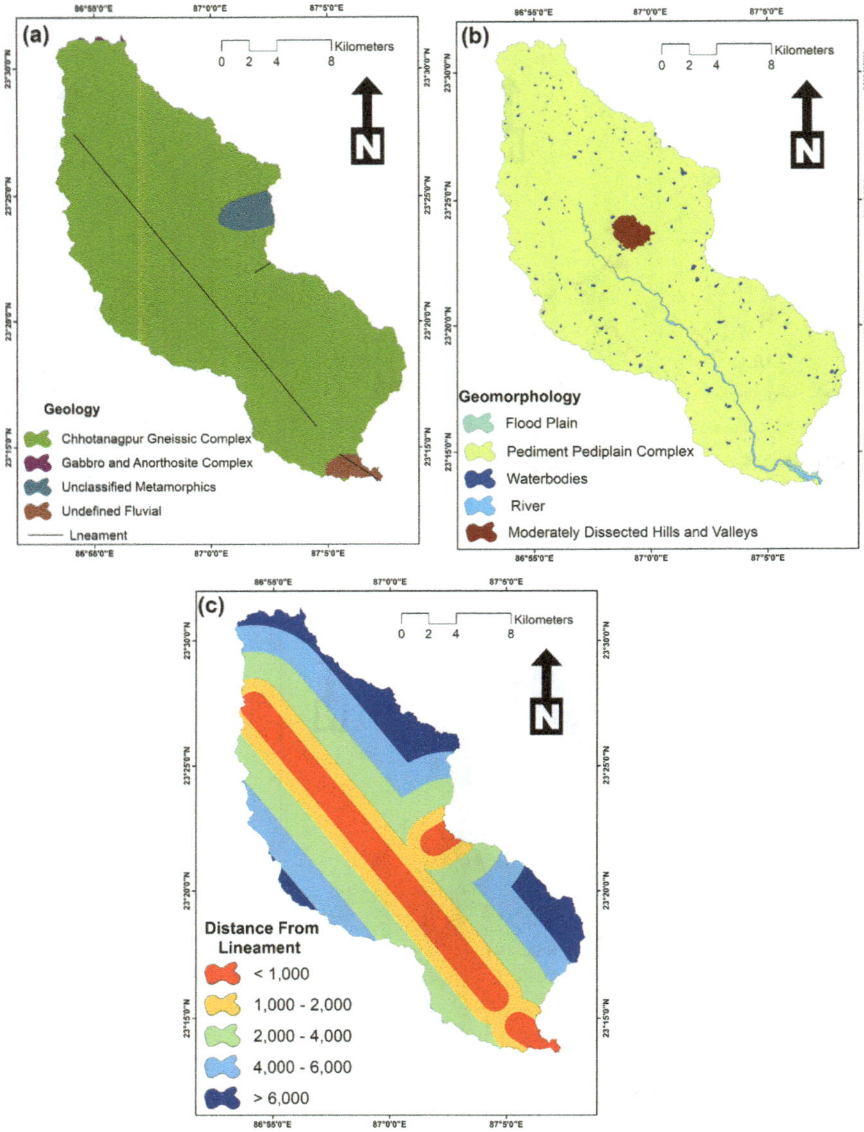

Fig. 6 Hydrological factors: Geology (**a**), geomorphology (**b**) and distance from lineament (**c**)

4.3.2 Multicolinearity Assessment

Linear regressions among the independent variables have been measured through Multicolinearity assessment [39]. The geo-environmental factors are correlated each other and there are some colinearity between the independent variables. Multicolinearity assessment was done through the VIF and tolerance (Eqs. 7 and 8). This index

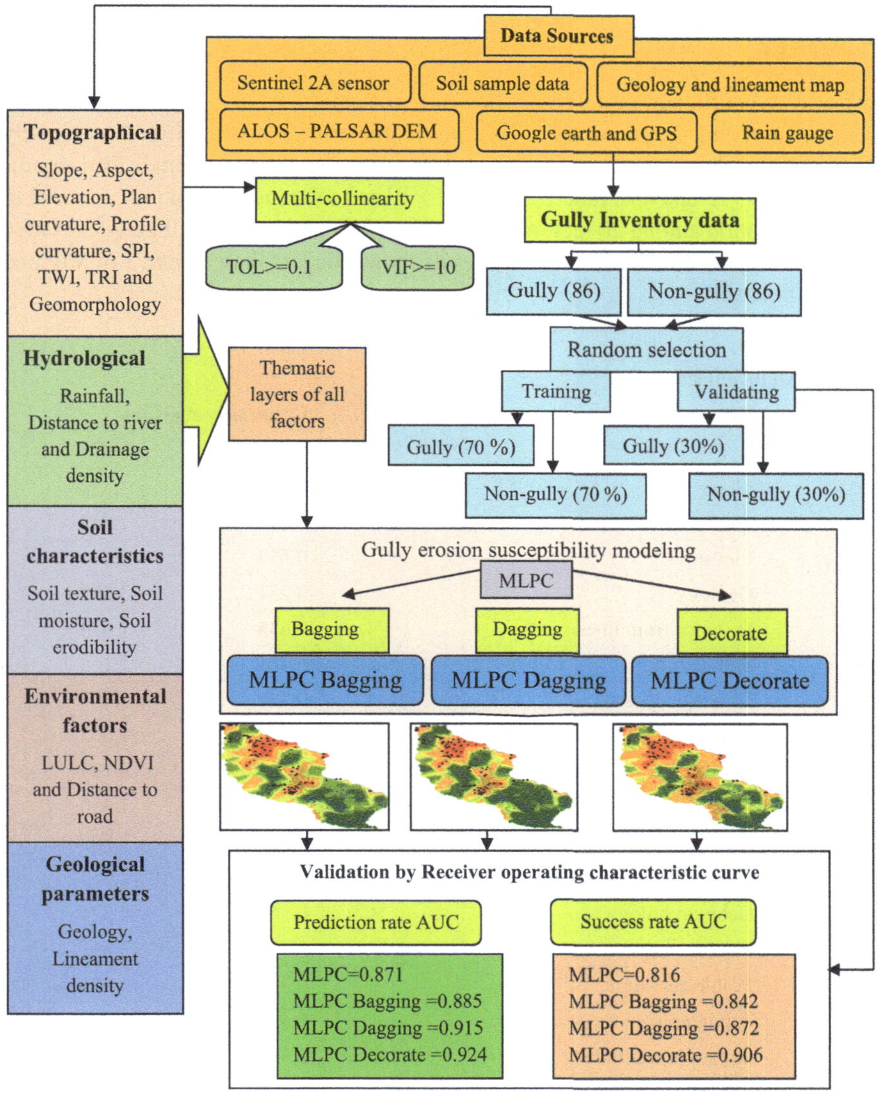

Fig. 7 Methodology flow chart

is used to show correlation between gully erosion occurring parameters. Menard [40], Hair et al. [41] has established where VIF value is below 10 and tolerance value is below 1 indicate perfect multi colinearity assessment.

$$\mathrm{TOL} = 1 - R^2 \tag{7}$$

$$\mathrm{VIF} = \frac{1}{\mathrm{TOL}} \quad (8)$$

where, R = coefficient of determination of regression.

To quantify factors there are many multicolinearity assessment method namely Pearson's correlation coefficient [42], Variance disintegration magnitude [43], conditional index, VIF and Tolerance [44]. Here VIF and tolerance techniques are used to show association among 20 factors. The result of VIF and TOL for all gully causative factors have shown in Table 1 and the result shows the less than 10 and 1 of VIF and TOL respectively.

Table 1 Multicollinearity test

No	Factors	Colinearity statistics	
		Tolerance	VIF
1	Slope	0.553	1.438
2	Slope aspect	0.518	1.948
3	Elevation	0.653	1.286
4	Geology	0.251	3.658
5	Distance from lineament	0.635	1.574
6	Soil texture	0.461	2.114
7	Drainage density	0.332	3.307
8	Distance from drainage	0.624	1.665
9	Rainfall	0.259	4.139
10	Distance from road	0.635	1.430
11	Land use and land cover	0.374	3.509
12	Normalized difference vegetation index	0.392	2.553
13	Geomorphology	0.268	4.321
14	Plan curvature	0.421	3.861
15	Profile curvature	0.385	3.201
16	Topographical wetness index	0.491	2.883
17	Stream power index	0.335	3.377
18	Terrain ruggedness index	0.684	1.695
19	Soil moisture	0.558	1.998
20	Soil erodibility	0.481	2.194

4.3.3 Approaches of Models for Gully Susceptible Map

MLPC (Multi-layer Perception Approach)

Multilayer perception classifier approach is a nonlinear statistical tool related with Artificial neural networks (ANN) which can make gully erosion susceptibility map. Before understand MLPC, we have to understand ANN first. ANN is a computing system, derived from biological neurons networks that are related to animal brains. This model performs operation by learning form examples rather than specifically programmed for any task-specific rules. ANN collected artificial neurons and these are similar to the neurons of animal brains. The connection of artificial neurons and synapses of any animal brain are same and the transform signals gone from one artificial neurons to another. ANN is used in different tasks such as speech recognition, social network filtering, machine translation, etc. The whole assessment was tested by three layers such as input layer, hidden layer, and output layer. It is also called as single-layer perception network consists of two layers namely input and output which are directly related to each other. The input layer is passive in character and receives primary data, where output and hidden layers are processing data. Finally output layers establish the neural network result. When inputs result indicates 1 then it takes activated route otherwise it takes deactivated route. This type of system has a relation to multilayer computational units and connected by fed forward way. Each and every neuron of layers has relationship to the neurons of subsequent layers. In the multi-layer perception approach output gives the correct result to calculate the value of a few pre-determined error of function. Numerous techniques uploaded the error back in the system. The algorithm adjusts each connection and reduces the value of error using different data, information and techniques. Repeating the procedure neuron network coverage the whole process and reduces the error and come to a state condition. Conditioning factors which are called as neurons considered in input and output layer and fixed by the application. Gong [45] has explained hidden neurons and determined trial and error.

MLPC is a popular machine learning programmed, has come to 1980s. It has a diverse application in diverse field such as speech reorganization and image reorganization. This system has as importance as Support vector machine (SVM). Cybenko's theorem [46] has said MLPC is a universal method which accurately establishes the result by programming models and ensembles approach. Bui et al. [47] has said that in MLPC model, relationship between input and output layer is complex. This model is a combination of input, output and hidden layer. Here, 20 factors are considered as neurons. Output has come as one neuron and training data determined hidden layer which is one or more than one. To increase the networks ability the model complexes functions, hidden layers play most important roles. These three layers have sufficient amount of neurons which depends on application.

Ensembles approaches of MLPC

Bagging
Breiman [48] has stated bootstrap aggregation is enabled through bagging. In spite of its straight forwardness and intuitive, bagging is such an algorithm which can describe the method. Such methods develop robustness against our layers. Bootstrap selection is done by this method. The expected error of bagging which is shown in theoretical result has a bias, as same as single bootstrap replicate, although variance is reduced. It also reduces the variance of categorization error. In bagging ensembles methods random subset are produced and a classifier is applied for each subset to provide a result.

Dagging
Dagging was developed by Ting and Witten [49] and it is a resampling ensembles technique. To improve predictive accuracy dagging use the majority vote for combination of diversity of classifiers [50]. Dagging uses disjoint samples where bagging uses bootstrap samples to produce base classifiers.

Decorate
By using artificially constructed training sample, decorate generate ensembles which directly constructs diverse classifiers [51]. To produce more accurate model, it is a larger ensembles. This method needs more complexity and also needs greater time to establish an accurate result [52]. Decorate is a framework which was first propounded by Melville [53]. For building the varied ensembles of classifiers, this technique specially constructed non-natural training instances. The form of Decorate has a similarity to the form of Adaboost. Diverse classifiers are built by modifying the learning data set. There is a differentiation between Decorate and Adaboost when they made training data set of each classifier. Decorate reduce correlation between ensembles algorithm. This method established that training error is less than or equivalent to the error of base algorithm [54].

4.4 Evaluation of Models

4.4.1 Receiver Operating Characteristic (ROC) Curve

Various methods were used to predict accuracy of gully erosion; ROC (Receiver operating characteristic) curve is one of them. This curve measures random point of presence and absence of gully in this area for measuring validation of gully erosion susceptible maps and here ROC curve measures the gully erosion susceptible map which was established through MLPC and its ensembles (Bagging, Dagging and Decorate). ROC represent 'X' and 'Y' co-ordinates to show true positive and false positive, respectively. Based on observed and predictive values ROC measures the accuracy. This curve shows specificity on 'X' axis and sensitivity on 'Y' axis [55,

56]. AUC values in ROC shows statistical summary of selected area where overall models result is used to authenticate and evaluate the models. Always validation has to be done for scientific result presentation. To evaluate scientific result, ROC curve with AUC approach is the best method [57, 58] has stated ROC is a binary classification of true positive and false positive dataset. The value of AUC ranges from 0 to 1. When it indicates 1 there will be no error. ROC curve accepts AUC value when it will be higher than 0.5 [59]. Whenever it shows 1, it performs best. AUC < 0.5 presents weak performances and AUC classify total result as poor (50–60%), moderate (60–70%), good (70–80%), very good (80–90%) and outstanding (90–100%).

4.4.2 Cohen Kappa's Model

Cohen kappa or kappa index is a method to measure the inter/intra-rater reliability for the qualitative items. LULC is the qualitative factors and classification of LULC is no cut of value. So if we do LULC classification of the study area then the area covered forest area, water bodies, agriculture land, builds up area, fallow land and shrubs area. And classify the study river basin based on field observation and from satellite image with the help Remote sensing (RS) and GIS environment. Here Cohen Kappa's model is best for measure the accuracy of classification (Eq. 9).

$$K = \frac{P_{\text{obs}} - P_{\text{exp}}}{1 - P_{\text{exp}}} \tag{9}$$

where, P_{obs} P is presents (TP + TN) correctly and, $P_{\text{exp}} = (\text{TP}+\text{FN}) + (\text{TP}+\text{FP}) + (\text{FP} + \text{TN})$

Landis and Koch [60] has shown Kappa's coefficient classify the model result into six categories like poor (≤ 0), fair (0.2–0.4), reasonable (0.4–0.6), considerable (0.6–0.8) and almost ideal (0.8–1). The Kappa index of LULC classification has shown the considerable value ($K = 0.75$).

5 Results and Discussion

5.1 Gully Erosion Susceptibility Assessment Using MLPC, Bagging-MLPC, Dagging-MLPC and Decorate-MLPC

The assessment of gully erosion susceptibility has been done with considering the hybrid computational approaches of different ensemble classifier. In MLPC, the very high gully erosion susceptible areas are found in the western, north-western and middle part. The high gully erosion susceptible areas are found in the western and middle part. The moderate gully erosion susceptible areas are found in the western, middle and southern part. The low gully erosion susceptible areas are mainly found in the middle and eastern part (Fig. 8a). The very low gully erosion susceptible areas are mainly found in the eastern part of the watershed. In MLPC-Bagging, the very high gully erosion susceptible areas are mostly found in the north-western and middle part. The high gully erosion susceptible areas are found in the western,

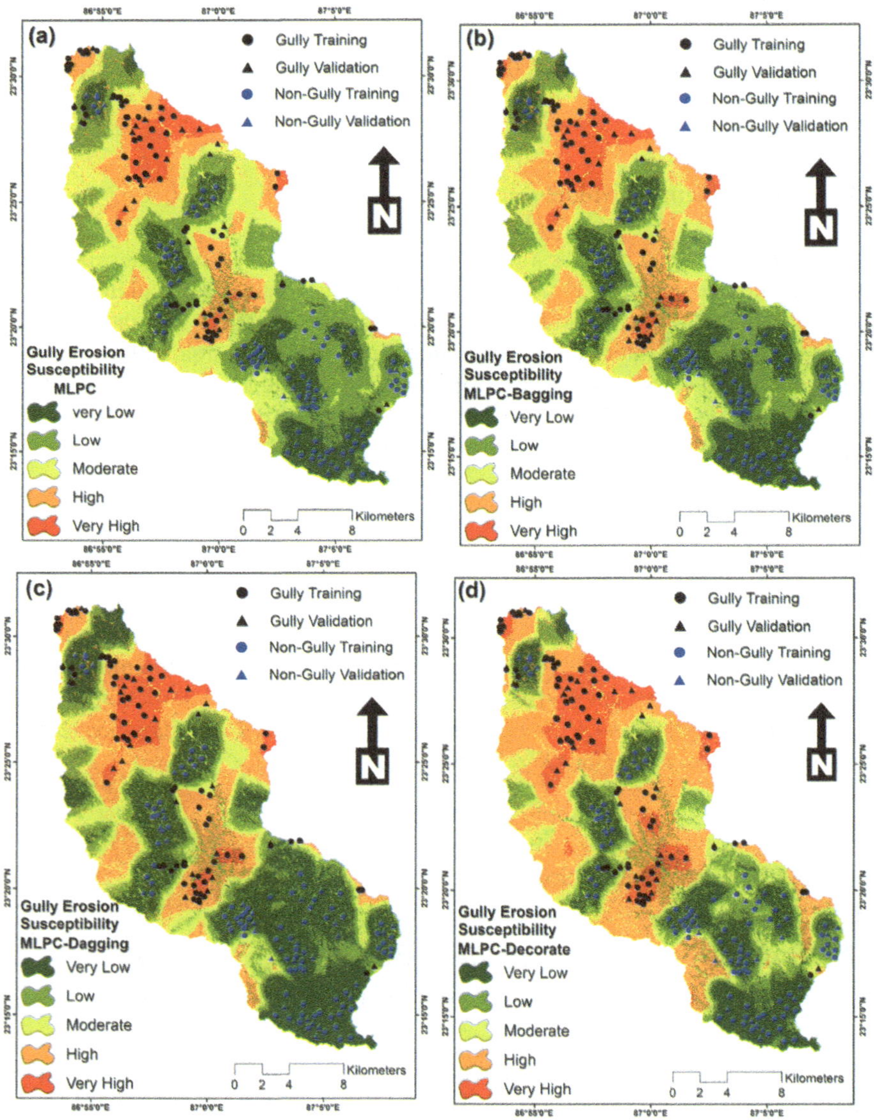

Fig. 8 Gully erosion susceptibility assessment using MLPC (**a**) MLPC-Bagging (**b**), MLPC-Dagging (**c**) and MLPC-Decorate (**d**)

north-western and middle part. The moderate gully erosion susceptible areas are found in the western, middle, north-eastern and southern part. The low gully erosion susceptible areas are mainly found in the western, middle and eastern part. The very low gully erosion susceptible areas are mostly found in the middle and eastern part of the watershed (Fig. 8b). In MLPC-Dagging, the very high gully erosion susceptible areas are mainly found in the north-western, western and middle part. The high gully erosion susceptible areas are found in the western, north-western, north-eastern and middle part. The moderate gully erosion susceptible areas are found in the western, middle and southern part. The low gully erosion susceptible areas are largely found in the western, middle and eastern part. The very low gully erosion susceptible areas are largely found in the middle and eastern part of the watershed (Fig. 8c). In MLPC-Decorate, the very high gully erosion susceptible areas are mostly found in the north-western and middle part. The high gully erosion susceptible areas are found in major portion of this region. The moderate gully erosion susceptible areas are found in the western, south-western, middle and eastern part. The low gully erosion susceptible areas are mostly found in the middle and eastern part. The very low gully erosion susceptible areas are largely found in the eastern part of the watershed (Fig. 8d).

Water-induced erosion in different forms of erosion is a frequent matter in the sub-tropical environment. The impact of storm rainfall event in monsoon period is responsible for this situation. Apart from this the association of long dry and short wet season is also responsible for large-scale erosion. The identification of gully erosion susceptible zone is very much essential to get the actual scenario and preventing for these environmental problems. In recent studies the application of multiple model and select the optimal model through scientific experiment is rapidly going on. The use of machine learning techniques in GIS platform has been considered as a reliable predictor which is capable to estimate the actual scenario with maximum possible accuracy. The selections of the influential parameters are one of the essential tasks for predicting the susceptible zones with adequate accuracy. From the multicolinearity analysis, it was found that, the all conditioning factors are associated with $VIF < 5$ and $TOL > 0.1$ (Table 1). These models are considered as a one of the reliable model which is deals less number of data and less effort.

5.2 *Validation*

For predict 30% gully inventory different models were Pourghasemi et al. [61] already has stated that there are various method for measuring accuracy of gully erosion and one of them important is Receiver operating characteristic (ROC) curve with AUC values. Dube et al. [62] has established, using as a geographical presentation AUC values perfectly indicates occurring gully area and non-occurring gully area. Being classified as five categories; as poor (0.5–0.6), moderate (0.6–0.7), good (0.7–0.8), very good (0.8–0.9) and excellent (1.0) AUC values with ROC predict gully erosion area and validate the model [63].

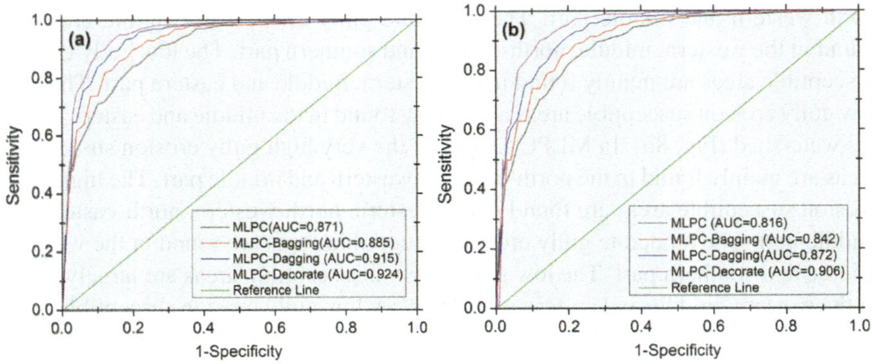

Fig. 9 Performances of the models using ROC; training (**a**) and validating (**b**)

There are extensive field information has been collected for finding out the actual scenario of land degradation and incorporated it for validation purpose (Fig. 10). The AUC of ROC for training datasets in MLPC, MLPC-Bagging, MLPC-Dagging, MLPC-Decorate are 0.871, 0.885, 0.915 and 0.924 respectively (Fig. 9a). The AUC of ROC for validating datasets in MLPC, MLPC-Bagging, MLPC-Dagging, MLPC-Decorate are 0.816, 0.842, 0.872 and 0.906 respectively (Fig. 9b). All the models are deals with high AUC values and good fitted with primary observation. The AUC values from ROC have been considered for identifying the optimal model.

6 Conclusion

The sub-tropical monsoon dominated region likewise Gandheswari watershed facing the extreme problems of land degradation due to several forms of erosion like sheet, rill and gully etc. (Fig. 10). The natural environment and local ecosystems are damages for extreme rate of gully erosion, which is generally the extreme form of land degradation. The formation of gullies can play a crucial function in damaging the surface layer and which is responsible for large-scale erosion. The newly constructed roads are also damages by origin and development of gullies. So the identification of gully erosion susceptible zones is very realistic for taking the suitable measures to escape this situation. In this study, MLPC and its different ensemble (MLPC-Bagging, MLPC-Dagging and MPPC-Decorate) methods are considered for estimating the susceptible areas. The work mainly encompasses to development the optimal model for susceptibility of gully erosion in sub-tropical area with considering the hybrid computational approaches. From this analysis it has been observed that, the ensemble classifier is much better than the separate machine learning algorithm. In ensemble classifier, the MLPC-Decorate has much better predator then the other ensemble classifier. The vast extensive field visit has been done in before and after monsoon period for accounting the actual impact of monsoonal rainfall and

Fig. 10 Primary field observation regarding the actual scenario of erosion in this region

validating purpose. Finally it can be said that, the MPLC-Decorate model can be used in sub-tropical region for any kind of susceptibility assessment without any modifications. Apart from this, the output results can be implemented in development strategies for reducing the risk of erosion. The local stake holders are the main beneficiary regarding this information, and they can take the traditional remedies with considering this outcome.

Acknowledgements We are grateful to the Department of Geography, The University of Burdwan for providing us the infrastructure to carry out the research work. We are also thankful to the anonymous Reviewers and the Editors of this book to provide valuable suggestion in regard to this work.

References

1. Zabihi M, Pourghasemi HR, Motevalli A, Zakeri MA (2019) Gully erosion modeling using GIS-based data mining techniques in Northern Iran: a comparison between boosted regression tree and multivariate adaptive regression spline. Natural hazards GIS-Based spatial modeling using data mining techniques. Springer, Cham, pp 1–26
2. Pal SC, Shit M (2017) Application of RUSLE model for soil loss estimation of Jaipanda watershed, West Bengal. Spat Inf Res 25(3):399–409
3. Arabameri A, Pradhan B, Rezaei K (2019) Gully erosion zonation mapping using integrated geographically weighted regression with certainty factor and random forest models in GIS. J Environ Manage 232:928–942
4. Mekonnen M, Keesstra SD, Baartman JE, Stroosnijder L, Maroulis J (2017) Reducing sediment connectivity through man-made and natural sediment sinks in the Minizr catchment, Northwest Ethiopia. Land Degrad Dev 28(2):708–717
5. Das H, Naik B, Behera HS (2018) Classification of diabetes mellitus disease (DMD): a data mining (DM) approach. In: Progress in computing, analytics and networking. Springer, Singapore, pp 539–549
6. Das H, Naik B, Behera HS (2019) Medical disease analysis using neuro-fuzzy with feature extraction model for classification. Inform Med Unlocked 100288
7. Das H, Naik B, Behera HS (2020) An experimental analysis of machine learning classification algorithms on biomedical data. In: Proceedings of the 2nd international conference on communication, devices and computing. Springer, Singapore, pp 525–539
8. Pal SC, Chowdhuri I (2019) GIS-based spatial prediction of landslide susceptibility using frequency ratio model of Lachung river basin, North Sikkim, India. SN Appl Sci 1(5):416
9. Das B, Pal SC (2019a) Assessment of groundwater recharge and its potential zone identification in groundwater-stressed Goghat-I block of Hugli District, West Bengal, India. Environ Dev Sustain 1–19
10. Das B, Pal SC (2019b) Combination of GIS and fuzzy-AHP for delineating groundwater recharge potential zones in the critical Goghat-II block of West Bengal, India. HydroResearch 2:21–30
11. Pal SC, Das B, Malik S (2019) Potential landslide vulnerability zonation using integrated analytic hierarchy process and GIS technique of Upper Rangit Catchment Area, West Sikkim, India. J Indian Soc Remote Sens 47(10):1643–1655
12. Das H, Jena AK, Nayak J, Naik B, Behera HS (2015) A novel PSO based back propagation learning-MLP (PSO-BP-MLP) for classification. In: Computational intelligence in data mining, vol 2. Springer, New Delhi, pp 461–471
13. Dey N, Ashour AS, Kalia H, Goswami R, Das H (2019a) Histopathological image analysis in medical decision making
14. Dey N, Das H, Naik B, Behera HS (eds) (2019b) Big data analytics for intelligent healthcare management. Academic Press
15. Rout M, Jena AK, Rout JK, Das H (2020) Teaching–learning optimization based cascaded low-complexity neural network model for exchange rates forecasting. In: Smart intelligent computing and applications. Springer, Singapore, pp 635–645
16. Sahani R, Rout C, Badajena JC, Jena AK, Das H (2018) Classification of intrusion detection using data mining techniques. In: Progress in computing, analytics and networking. Springer, Singapore, pp 753–764
17. Sahoo AK, Pradhan C, Das H (2020) Performance evaluation of different machine learning methods and deep-learning based convolutional neural network for health decision making. Nature inspired computing for data science. Springer, Cham, pp 201–212
18. Craglia M, Onsrud H (2014) Geographic information research: transatlantic perspectives. CRC Press
19. Akgün A, Türk N (2011) Mapping erosion susceptibility by a multivariate statistical method: a case study from the Ayvalık region, NW Turkey. Comput Geosci 37(9):1515–1524

20. Chowdhuri I, Pal SC, Chakrabortty R (2019) Flood susceptibility mapping by ensemble evidential belief function and binomial logistic regression model on river basin of eastern India. Adv Space Res
21. Lombardi M, Milano M (2018) Boosting combinatorial problem modeling with machine learning. arXiv preprint arXiv:1807.05517
22. Griggs D, Stafford-Smith M, Gaffney O, Rockström J, Öhman MC, Shyamsundar P, Steffen W, Glaser G, Kanie N, Noble I (2013) Policy: sustainable development goals for people and planet. Nature 495(7441):305
23. Ligonja PJ, Shrestha RP (2015) Soil erosion assessment in kondoa eroded area in Tanzania using universal soil loss equation, geographic information systems and socioeconomic approach. Land Degrad Dev 26(4):367–379
24. Pal SC, Chakrabortty R (2018) Modeling of water induced surface soil erosion and the potential risk zone prediction in a sub-tropical watershed of Eastern India. Model Earth Syst Environ 5(2):369–393
25. Pal SC, Chakrabortty R (2019) Simulating the impact of climate change on soil erosion in sub-tropical monsoon dominated watershed based on RUSLE, SCS runoff and MIROC5 climatic model. Adv Space Res 64(2):352–377
26. Chakrabortty R, Ghosh S, Pal SC, Das B, Malik S (2018) Morphometric analysis for hydrological assessment using remote sensing and GIS technique: a case study of Dwarkeswar river basin of Bankura district, West Bengal. Asian J Res Soc Sci Humlties 8(4):113–142
27. EIA-EMP-Report (2007) Irrigation and Waterways Directorate, Government of West Bengal, India
28. Arabameri A, Pradhan B, Rezaei K, Yamani M, Pourghasemi HR, Lombardo L (2018) Spatial modelling of gully erosion using evidential belief function, logistic regression, and a new ensemble of evidential belief function–logistic regression algorithm. Land Degrad Dev 29(11):4035–4049
29. Conoscenti C, Angileri S, Cappadonia C, Rotigliano E, Agnesi V, Märker M (2014) Gully erosion susceptibility assessment by means of GIS-based logistic regression: a case of Sicily (Italy). Geomorphology 204:399–411
30. Gascon F, Bouzinac C, Thépaut O, Jung M, Francesconi B, Louis J, Lonjou V, Lafrance B, Massera S, Gaudel-Vacaresse A Languille F (2017) Copernicus Sentinel-2A calibration and products validation status. Remote Sens 9(6):584
31. Pal SC, Chakrabortty R, Malik S, Das B (2018) Application of forest canopy density model for forest cover mapping using LISS-IV satellite data: a case study of Sali watershed, West Bengal. Model Earth Syst Environ 4(2):853–865
32. Shimabukuro YE, Duarte V, Arai E, Freitas RM, Lima A, Valeriano DM, Brown IF, Maldonado MLR (2009). Fraction image segmentation to evaluate deforestation in Landsat Thematic Mapper images of the Amazon region. Int J Remote Sens 19(3):535–541
33. Teng H, Liang Z, Chen S, Liu Y, Rossel RAV, Chappell A, Yu W, Shi Z (2018) Current and future assessments of soil erosion by water on the Tibetan Plateau based on RUSLE and CMIP5 climate models. Sci Total Environ 635:673–686
34. Rahmati O, Tahmasebipour N, Haghizadeh A, Pourghasemi HR, Feizizadeh B (2017) Evaluation of different machine learning models for predicting and mapping the susceptibility of gully erosion. Geomorphology 298:118–137
35. Das B, Pal SC, Malik S, Chakrabortty R (2019) Living with floods through geospatial approach: a case study of Arambag CD block of Hugli district, West Bengal, India. SN Appl Sci 1(4):329
36. Das B, Pal SC, Malik S, Chakrabortty R (2019) Modeling groundwater potential zones of Puruliya district, West Bengal, India using remote sensing and GIS techniques. Geol Ecol Landsc 3(3):223–237
37. Malik S, Pal SC, Das B, Chakrabortty R (2019a) Assessment of vegetation status of Sali River basin, a tributary of Damodar river in Bankura district, West Bengal, using satellite data. Environ Dev Sustain 1–35
38. Malik S, Pal SC, Das B, Chakrabortty R (2019b) Intra-annual variations of vegetation status in a sub-tropical deciduous forest-dominated area using geospatial approach: a case study of Sali watershed, Bankura, West Bengal, India. Geol, Ecol, Landsc, 1–12

39. Alin A (2010) Multicollinearity. Wiley Interdiscip Rev: Comput Stat 2(3):370–374
40. Menard S (2002) Applied logistic regression analysis, vol 106. Sage
41. Hair B, Black WC, Babin B, Anderson RE (2006) Tatham, multivariate data analysis
42. Sedgwick P (2012) Pearson's correlation coefficient. BMJ 345:e4483
43. Schuerman JR (2012) Multivariate analysis in the human services, vol 2. Springer Science & Business Media
44. Dormann CF, Elith J, Bacher S, Buchmann C, Carl G, Carré G, Marquéz JRG, Gruber B, Lafourcade B, Leitão PJ, Münkemüller T (2013) Collinearity: a review of methods to deal with it and a simulation study evaluating their performance. Ecography 36(1):27–46
45. Gong P (1994) Integrated analysis of spatial data from multiple sources: an overview. Can J Remote Sens 20(4):349–359
46. Cybenko G (1989) Approximation by superpositions of a sigmoidal function. Math Control Signals Syst 2(4):303–314
47. Bui DT, Pradhan B, Lofman O, Revhaug I, Dick OB (2012) Landslide susceptibility mapping at Hoa Binh province (Vietnam) using an adaptive neuro-fuzzy inference system and GIS. Comput Geosci 45:199–211
48. Breiman L (1996) Bagging predictors. Mach Learn 24(2):123–140
49. Ting KM, Witten IH (1997) Stacking bagged and dagged models
50. Kotsianti SB, Kanellopoulos D (2007) Combining bagging, boosting and dagging for classification problems. In: International conference on knowledge-based and intelligent information and engineering systems. Springer, Berlin, Heidelberg, pp 493–500
51. Melville P, Mooney RJ (2003) Constructing diverse classifier ensembles using artificial training examples. IJCAI 3:505–510
52. Witten IH, Frank E, Hall MA (2011) Data mining: practical machine learning tools and techniques 3
53. Melville P (2003) Creating diverse ensemble classifiers. Computer Science Department, University of Texas at Austin, p 34
54. Rodriguez JJ, Kuncheva LI, Alonso CJ (2006) Rotation forest: a new classifier ensemble method. IEEE Trans Pattern Anal Mach Intell 28(10):1619–1630
55. Pham BT, Bui DT, Prakash I, Nguyen LH, Dholakia MB (2017) A comparative study of sequential minimal optimization-based support vector machines, vote feature intervals, and logistic regression in landslide susceptibility assessment using GIS. Environ Earth Sci 76(10):371
56. Pham BT, Khosravi K, Prakash I (2017) Application and comparison of decision tree-based machine learning methods in landside susceptibility assessment at Pauri Garhwal Area, Uttarakhand, India. Environmental Processes 4(3):711–730
57. Hong H, Pradhan B, Xu C, Bui DT (2015) Spatial prediction of landslide hazard at the Yihuang area (China) using two-class kernel logistic regression, alternating decision tree and support vector machines. CATENA 133:266–281
58. Zweig MH, Campbell G (1993) Receiver-operating characteristic (ROC) plots: a fundamental evaluation tool in clinical medicine. Clin Chem 39(4):561–577
59. Yesilnacar E, Topal T (2005) Landslide susceptibility mapping: a comparison of logistic regression and neural networks methods in a medium scale study, Hendek region (Turkey). Eng Geol 79(3–4):251–266
60. Landis JR, Koch GG (1977) An application of hierarchical kappa-type statistics in the assessment of majority agreement among multiple observers. Biometrics 363–374
61. Pourghasemi HR, Yousefi S, Kornejady A, Cerdà A (2017) Performance assessment of individual and ensemble data-mining techniques for gully erosion modeling. Sci Total Environ 609:764–775
62. Dube F, Nhapi I, Murwira A, Gumindoga W, Goldin J, Mashauri DA (2014) Potential of weight of evidence modelling for gully erosion hazard assessment in Mbire District-Zimbabwe. Phys Chem Earth Parts A/B/C 67:145–152
63. Yesilnacar EK (2005) The application of computational intelligence to landslide susceptibility mapping in Turkey. PhD thesis, Department of Geomatics, The University of Melbourne, p 423

Chapter 2
Classification of ECG Heartbeat Using Deep Convolutional Neural Network

Saroj Kumar Pandey, Rekh Ram Janghel, and Kshitiz Varma

Abstract The report of World Health Organization (WHO) specifies that the diagnosis and treatment of cardiovascular diseases are challenging tasks. To study the electrical conductivity of the heart, Electrocardiogram (ECG) which is an inexpensive diagnostic tool, is used. Classification is the most well-known topic for arrhythmia detection related to cardiovascular disease. Many algorithms have been evolved for the classification of heartbeat arrhythmia in the previous few decades using the CAD system. In this paper, we have developed a new deep CNN (11-layer) model for automatically classifying ECG heartbeats into five different groups according to the ANSI-AAMI standard (1998) without using feature extraction and selection techniques. The experiment is performed on publicly available Physionet MIT-BIH database and evaluated results are then compared with the existing works mentioned in the literature. To handle the problem of minority classes as well as the class imbalance problem, the database has been oversampled artificially using SMOTE technique. The augmented ECG database was employed for training the model while the testing was performed on the unseen dataset. On evaluation of the results from the experiment, we found that the proposed CNN model performed better in comparison to the experiments mentioned in other papers in terms of accuracy, sensitivity, and specificity. *abstract* environment.

Keywords CNN · Arrhythmia · ECG signal · Class imbalance · Classification

---

S. K. Pandey (✉) · R. R. Janghel
Department of Information Technology, NIT Raipur, Chhattisgarh, India
e-mail: skpandey.phd2016.it@nitrr.ac.in

R. R. Janghel
e-mail: rrjanghel.it@nitrr.ac.in

K. Varma
CSVTU Bhilai, Durg, India
e-mail: vermaksh@gmail.com

J. K. Rout et al. (eds.), *Machine Learning for Intelligent Decision Science*,
Algorithms for Intelligent Systems, https://doi.org/10.1007/978-981-15-3689-2_2

1 Introduction

Millions, worldwide, are suffering from cardiovascular disease [1]. In accordance with the WHO report, 30% of deaths in the world are due to cardiovascular disease [2]. The death of cardiovascular disease is most commonly caused by cardiac death, the main cause of which is cardiac arrhythmia. When the functioning of our heartbeat becomes abnormal, it represents the cardiac arrhythmia, which is related to heart diseases [3, 4].

Due to arrhythmia, the heart rhythm is erratic, i.e., either it is too fast or too slow as compared to the regular rhythm of the heart, henceforth the heart finds it difficult to pump blood to different parts of the body in the correct proportion which significantly affects the heart, brain, and other major organs of the human body. Majorly arrhythmias cause a lot of harm to the heart and they are mainly categorized as life-threatening arrhythmias and non-life-threatening arrhythmias. Hence, continuous monitoring of heartbeat activity is just not possible. In order to deliver proper medical assistance, it is very critical to ascertain the arrhythmia [5–7]. An electrocardiogram (ECG) is by far the best tool for monitoring as well as identifying arrhythmia. To study the electrical conductivity of the heart, Electrocardiogram (ECG) which is an inexpensive diagnostic tool, is used [8]. Figure 1 shows the P wave, the QRS complex, and the T wave repeating in a periodic fashion.

A U wave is rarely detected [10]. The main feature of the ECG signal is the Q-R-S wave. A convenient ECG capturer named as Holter monitor is a device generally used to record longer duration ECG signals. However, manually beat by beat long duration ECG signal examination is very tedious and exhausting work to healthcare specialist. Therefore, clinicians generally employ computer-aided diagnosis methods to investigate ECG data and identification of arrhythmia, because of its robustness and effectiveness. Arrhythmia identification follows the detection of successive ECG beat categories in the given data. Hence classification of the heartbeat signal is an important step to recognize arrhythmia [11].

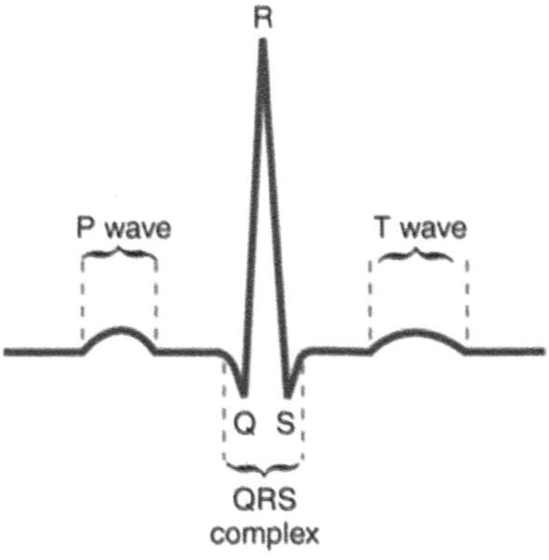

Fig. 1 Normal electrocardiogram signal [4]

1.1 State of the Art

Several techniques have been evolved for the classification of ECG arrhythmia in the past few years [12–18] using the CAD system. Basically, CAD classification of heartbeat arrhythmia consists of three parts: ECG preprocessing, feature extraction, and classification. The second and third phase is the most important part of arrhythmia classification, which has been described in many research papers. The morphological features can be extracted in two major ways, either directly from the electrocardiogram data or after applying a transformation method. The most used features are RR interval, PR interval, P duration, QRS duration, T duration, and used QT intervals [19]. These characteristics have been studied clinically and the associated diagnostic has been stipulated.

Moreover, frequency characteristics study has also contributed considerable appreciation into heartbeat signal processing. Signal processing techniques contain principal component analysis, wavelet decomposition, and independent component analysis. Although these techniques have an associated mathematical interpretation which makes it hard for doctors to comprehend. Therefore, to improve the classification performance, it may be a big challenge to choose the most effective features in the group of possible features [20–24].

In the third part usual classification algorithms such as Multilayers Perceptron [25], Random Forest [26], Bayesian Network [27], Support Vector Machine [28], etc. have been applied to get the positive class of the test samples. There exists an extra challenge associated with healthcare data named as class-imbalanced data. This challenge occurs because of the very less amount of data available in some classes named as minority classes which affects by the distribution of abnormal data. In the MIT-BIH ECG database, the outcomes will be biased near major class. Thus, it is a challenging task for supervised learning to decrease the misdiagnosis rate [29, 30]. Thus, very few research papers have considered class imbalance problem in their classification phase.

1.2 Contribution

The most studies related to ECG arrhythmia classification generally focused on the traditional machine learning (ML) perspectives like preprocessing feature selection and classification. However, these techniques demonstrated convenient ECG arrhythmia classification performance. Whenever CAD models are designed using traditional machine learning techniques, it generally faces overfitting problem in the testing phase and exhibits poor performance when validation is performed on a separate database [31].

Deep learning techniques have better capability to self-learn valuable features compared to unlike conventional techniques from ECG signals [32]. Therefore, with the help of these ideas, we have developed a new deep learning-based convolutional neural network (CNN) model to increase the performance of automatically arrhythmia classification by using segmentation of the input ECG signals [33]. We have compared our model with the existing classifiers mentioned in the paper. On analysis, our model outperforms the common classifiers. In this paper, we also performed some sampling method to make uniform data such as synthetic minority oversampling techniques, and random sampling to perceive the influence of class imbalance on the training data and testing data [34].

Finally, in this work, a deep learning CNN model is used to detect the five classes of ECG heartbeat. Rest of the paper is organized as follows: Sect. 2 discusses the ECG database and overview of theoretical background. Section 3 presents the main method used in this paper involving architecture. Section 4 discusses the experimental results and comparison with other existing works. Conclusion is mentioned in Sect. 5.

2 Database Used

In this work, the ECG signals have been taken from the publically available and popular MIT-BIH arrhythmia database [35]. This dataset is a standard test dataset applied since 1981 in numerous scientific tasks in the assessment of arrhythmia detection and classifications. The above dataset was combined from 48 electrocardiogram recordings, which were in turn taken from 47 subjects. The subjects included Twenty-two women, the range of whose age varied from 23 to 89 and Twenty-five men where the range varied from 32 to 89.

Each signal file in the arrhythmia database consists of the data generated from different electrodes placed on patient's chest. The data is available as a combination of two leads. Maximum signal files have contained modified limb lead II(MLII) and

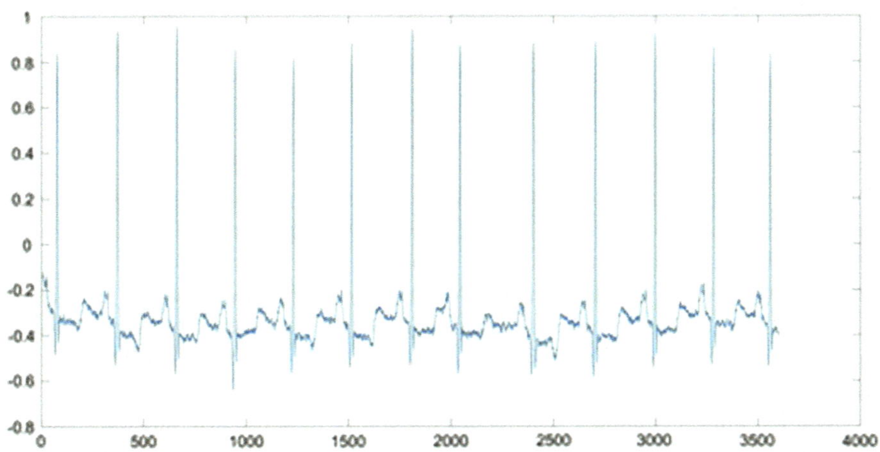

Fig. 2 MIT-BIH arrhythmia database extracted signal

Table 1 Relationship between AAMI standards and MIT-BIH arrhythmia database

ANSI-AAMI Standard heartbeat	MIT-BIH database heartbeats
Normal (N)	Normal, Left bundle branch block, Right bundle branch block Atrial escape, Nodal escape
Supra ventricular ectopic beat (S)	Aberrated aterial premature, Supraventricular premature, Atrial premature, Contraction nodal premature
Ventricular ectopic beat (V) Premature ventricular contraction	Ventricular escape
Fusion beat (F)	Fusion of ventricular and normal beat
Unclassified beat (Q) unclassifiable	Paced, Fusion of paced and normal

V1 recorded data. For example, a record of the arrhythmia database is extracted and depicted in Fig. 2.

These signals were obtained by using a bandpass filter at frequency 0.1 to 100 Hz and sampling rate at 360 Hz. There is also a reference file for each signal file in the MIT-BIH arrhythmia database, which is referred to as the heartbeat class and its position. Here, heartbeat class's annotation is taken as the standard notation for the detected heartbeat categories [36]. In accordance with ANSI/AAMI standards, four records, namely, 102, 104, 107, and 217, contain paced beats. Hence these records are discarded for evaluation purposes. The heartbeats left are then combined into 15 original classes and rearranged into 5 major categories as per the ANSI/AAMI standards. The relationship between AAMI standard heartbeat categories and MIT-BIH arrhythmia database classes is mentioned in Table 1 [37].

3 Methodology

The proposed arrhythmia classification method is split into three segments named as database normalization, heartbeat segmentation, and classification. Figure 3 summarizes the above three segments. Our main contribution is in the classification segment.

3.1 Arrhythmia Database Normalization

The electrocardiogram signal values here are normalized using the very popular normalization method known as the Z-score method. Problems such as amplitude scaling as well as vanishing the offset effects are easily resolved using this method. The formula used for normalization is as follows [38].

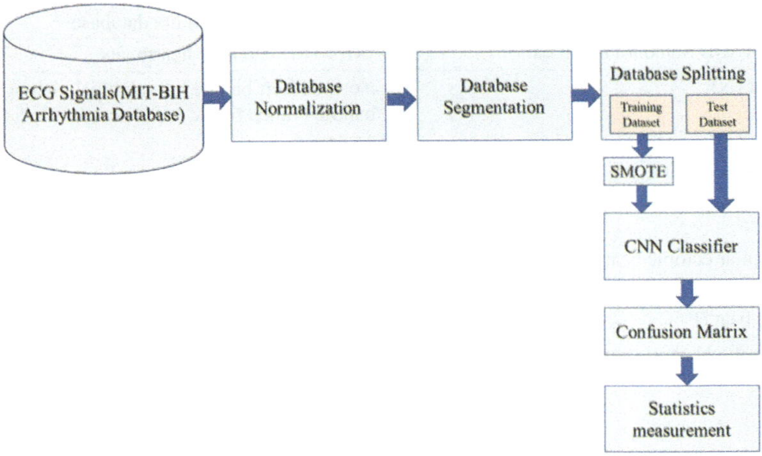

Fig. 3 Proposed arrhythmia classification methodology

$$\text{signal values} = \frac{\text{signal values} - \text{mean of signal values}}{\text{standard deviation of signal values}} \tag{1}$$

3.2 Heartbeat Segmentation

The MIT-BIH arrhythmia database is nothing but a record of continuous ECG signals. To retrieve an individual's heartbeat, the signals from electrocardiogram are stored after performing segmentation in the annotation files. The segmentation process involves finding the peak values from annotation files and then segmenting the ECG signals into segments based on those peaks. The length of each segment is 360 and is centered around the R peak positions [39].

3.3 Class Imbalance to Balance by Artificial Data Generation

The MIT-BIH arrhythmia database is very irregular and has a non-uniform heartbeat distribution of beats into various classes. As per the ANSI/AAMI standards, more than 80% of heartbeats belong to a single class while the other 20% belong to four different classes. Hence, this dataset can be easily categorized as an imbalanced dataset problem because of the fact that the sample number in the majority class is way too high as compared to the sample number in the minority class.

In the ECG database, class imbalance problem is an important issue and this problem is also experienced by many other domains like medical diagnosis [40] bio-informatics [41], text classification [42], etc. This affects the performance measurement of the standard classification algorithms, which assumes that the classes are balanced and data are uniformity distributed in the class. In recent times researchers are also spending appropriate attention to such problems.

Various approaches like Aagorithm level, data level, and hybrid of both are used for solving the class imbalance problem. In this proposed paper, two data level approaches are used, namely, Resampling method and Synthetic Minority Oversampling Technique (SMOTE).

3.3.1 Resampling Method

In this method, data distribution in the classes is balanced by using two strategies namely as random under sampling and oversampling [43]. In random under sampling strategies generates sub instances of the majority class are generated while in the random oversampling strategies the minority class samples are increased via random replication of minority class instances.

3.3.2 Synthetic Minority Oversampling Technique (SMOTE)

Introduced by Chawla in [44], it is one of the most widely used oversampling techniques. In this method, new synthetic minority samples get generated without any duplication in the samples. The generation is done by employing the following equation.

$$x^{\text{synthetic}} = x^i + (x^j - x^i) \times \delta \tag{2}$$

where x^i denotes the instance of minority class, x^j is an instance which is selected randomly from minority instance x^i K-nearest neighbors. δ is nothing but a vector where in every entity is a number in the range [0,1]; this in turn helps to produce a synthetic instance with reference to primary reference instance, x^i and assistant reference instance, x^j, respectively. $x^{\text{synthetic}}$ is produced along the line starting from x^i and stretching to x^j whose direction is toward the line joining x^i. This greatly helps in reducing the risk due to over fitting. With the above procedure, we get a synthetically created balancing set, where the training database of each class has equal number of instances. In accordance with ANSI/AAMI standards, electrocardiogram signal dataset is divided into five classes. The four classes' samples other than the normal class, namely, (S, V, F, and Q) are oversampled in order to remove the class imbalance problem [43] (Table 2).

3.4 Convolutional Neural Network (CNN)

CNN is a very popular kind of MLP which is made up of artificial neurons having weights, biases, and activation functions that map the input layer to the output layer

Table 2 Table depicting the total heartbeats for training database before and after oversampling method

Heartbeat type	Before artificial data generation	After artificial data generation using 90% training dataset
N	87542	78840
S	2726	78840
V	7221	78840
F	802	78840
Q	3888	78840
Total	102179	394200

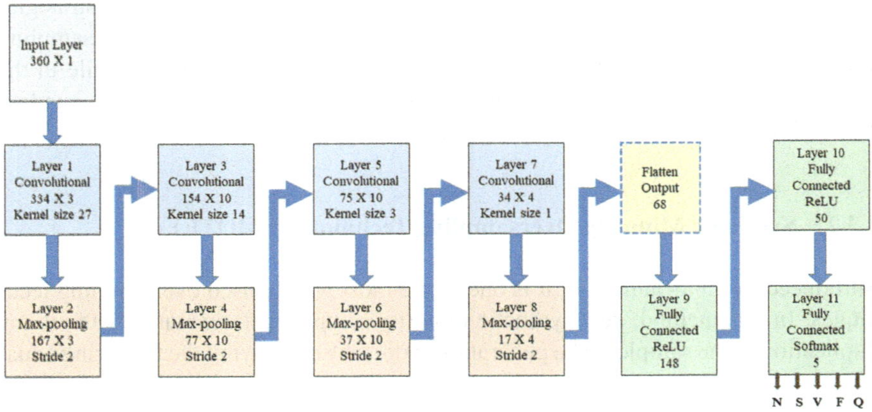

Fig. 4 Architecture of proposed CNN model for classification

[45]. In MLP, when the number of hidden layers is increased then it is called deep MLP. Similarly, CNN has also been considered as a special neural network of deep MLP. This unique structure enables the CNN model for both translation and rotation invariant [46]. The CNN architecture consists of three primary layers, namely, convolutional layer, pooling layer, and fully-connected layer topped with activation function.

3.4.1 Architecture of Proposed CNN Model

The proposed CNN model consists of 11 layers. There are four convolutional layers, four max-pooling layers, and three fully-connected layers as shown in Fig. 4.

The most outstanding feature of the architecture of CNN is that it can by itself extract and emphasize the most important and essential features from the input data by employing convolution operation. In CNN, more focus and concentration is spent on the local features and its position among other features. A lot of learning time is saved by network learning in parallel which is achieved by keeping the weights of neurons the same [47]. In this work, four convolution layers, namely, the first, third, fifth, and seventh having a filter size of 27, 14, 3, and 1, respectively, are convolved as per Eq. 2.

$$Y_n = \sum_{i=0}^{N-1} x_i f_{n-i} \tag{3}$$

In the above equation, the output vector is represented by Y, the signal values are denoted by x, the filters are denoted by f while N stands for the number of elements. A second max-polling layer is applied after each convolution layer to reduce the size of the feature map. In the layers 1, 3, 5, 7, 9, and 10, for the obtained parameter of the corresponding size, Brute force approach is employed. Activation function used is rectifier Linear Unit (ReLU) [48]. The network has three fully-connected layers having output neurons with size 148, 50, and 5, respectively. To accurately classify all the output classes (as N, S, V, F, & Q), soft-max function is used in the final layer (11). In this discussed model, CNN was trained by employing a backpropagation algorithm and using Adam optimizer technique. For proper training of the model, following parameters are used learning rates are 1e − 4, 1e − 3, and regularization of 0.01, respectively. The batch size in this CNN model is 34. The categorical cross-entropy function is applied for calculating the loss parameter. The training and testing are then performed on this CNN model in an iterative fashion. The iteration count is set as 50 and validation is performed after every. In this algorithm, firstly the real data is divided randomly in to 10 equal sections, i.e., 10% of data in each section. In the first section, we considered 10% real data for validation and the remaining 90 % data is oversampled using sampling techniques, then this sampled data is given to the CNN model for training. Similarly in second section, 10% data is taken as the next real data for validation and remaining data is oversampled and then trained with CNN model. This process is repeated ten times by using different sections of real data for validation. Finally, the statistical performance is measured by taking the average performance of all ten records.

4 Experimental Results

The presented CNN model was trained and tested on a Dell workstation having dual Intel Xeon E5-2600, 2.4 GHz as the processor with 64GB RAM. The average CPU times for learning rate $1e - 4$ is 486 seconds per epochs and for learning rate $1e - 3$ is 496 seconds per epochs for performing the training and testing on the database. This CNN model was developed in python using Keras libraries.

Performance metrics for the model were accuracy (Acc), Senstivity (Se), Specificity, precision, and F1-scores. The above are calculated with the help of true positive (TP), false positive (FP), false negative (FN), true negative (TN), and the various number of instances depicted and presented ($\sum$) are as follows:

$$\text{Overall Accuracy} = \frac{\text{TP}_1 + \text{TP}_2 + \ldots + \text{TP}_\text{N}}{\sum} \tag{4}$$

$$\text{Accuracy(Binary} - \text{class)} = \frac{\text{TP} + \text{TN}}{\text{TP} + \text{FN} + \text{TN} + \text{FP}} \tag{5}$$

$$\text{Recall}(Se) = \frac{\text{TP}}{\text{TP} + \text{FN}} \tag{6}$$

$$\text{Specificity } (Sp) = \frac{\text{TN}}{\text{TN} + \text{FP}} \tag{7}$$

$$\text{Precision } (P) = \frac{\text{TP}}{\text{TP} + \text{FP}} \tag{8}$$

$$\text{F-score} = 2\frac{P * Se}{P + Se} \tag{9}$$

The confusion matrix of the CNN model for detailed analysis of above-mentioned formulas is shown in Tables 3, 4, 5, and 6. Using the proposed CNN model, it maps the right and wrong identification of heart signals. In this confusion matrix, the original truth is provided by the annotation file in the database. The confusion matrix is represented by the rows and columns portray detected ECG signal classified by our proposed model. In Tables 3, 4, 5, and 6, the information of statistical calculation parameters for all ECG heartbeat classes has been reported according to Eqs. 4–8. In this table the details of classification results, recall, specificity, precision, F1-score, and accuracy are calculated for each class based on analysis of confusion matrix from tables.

The overall accuracy of the proposed CNN model is calculated using the Eq. 3, which is given in the following Table 7.

The training accuracy, validation accuracy, and training loss and validation loss using different learning rates and different sampling methods are shown in the following Figs. 5, 6, 7, 8, 9, 10, 11, and 12.

The performance comparison of classification results in terms of accuracy between the proposed CNN model and other existing methods in literature from [18, 50–57] is given in Table 8. The proposed model performed well in heartbeat signal classification accuracy compared with the existing works.

KNN (K-nearest neighbor), HOSF (Higher order statistic function), MF (Morphological feature), FFT (Fast Fourier transform), SVM (Support vector machine), PCA (Principal component analysis), HOS (Higher order statistics), DT (Decision tree), PNN (Probabilistic neural network), WPE (Wavelet packet entropy), RS-QNN (Rough set-Quantum neural network).

Table 3 Average of all tenfold ECG heartbeat confusion matrix and statistical performance for SMOTE + learning rate 1e-4 on testing dataset

Original	Predicted										
	HBT	N	S	V	F	Q	Recall (%)	Specificity (%)	Precision	F-score	Accuracy (%)
	N	8579	62	21	39	1	98.58	97.75	99.6	99.09	98.46
	S	15	261	4	1	0	92.88	99.33	79.57	85.71	99.15
	V	13	5	694	8	0	96.38	99.67	95.72	96.05	99.44
	F	5	0	4	81	0	90.00	99.53	62.79	73.79	99.44
	Q	1	0	2	0	418	99.28	99.99	99.76	99.52	99.96

Table 4 Average of all tenfold ECG heartbeat confusion matrix and statistical performance for SMOTE + learning rate 1e-3 on testing dataset

Original	Predicted									
	HBT	N	S	V	F	Q	Recall (%)	Specificity (%)	Precision (%)	F-score (%)	Accuracy (%)
	N	8470	142	26	58	6	97.33	97.75	99.55	98.43	97.36
	S	16	261	3	1	0	92.88	98.51	63.81	75.65	98.36
	V	15	6	690	9	0	95.83	99.63	95.17	95.50	99.36
	F	6	0	4	80	0	88.89	99.33	54.04	67.23	99.24
	Q	1	0	2	0	418	99.29	99.94	98.58	99.93	99.91

Table 5 Average of all tenfold ECG heartbeat confusion matrix and statistical performance for Oversampling + learning rate 1e-4 on testing dataset

Original	Predicted										
	HBT	N	S	V	F	Q	Recall (%)	Specificity (%)	Precision (%)	F-score (%)	Accuracy (%)
	N	8493	116	52	38	3	97.60	98.74	99.78	98.68	97.77
	S	10	267	2	1	1	95.02	98.77	68.64	79.70	98.67
	V	5	6	701	8	0	97.36	99.29	91.28	94.22	99.16
	F	3	0	11	76	0	84.44	99.54	61.79	71.36	99.40
	Q	1	0	2	0	418	99.29	99.96	99.05	99.17	99.93

Table 6 Average of all tenfold ECG heartbeat confusion matrix for Oversampling + learning rate 1e-3 on testing dataset

Original	Predicted										
	HBT	N	S	V	F	Q	Recall (%)	Specificity (%)	Precision (%)	F-score (%)	Accuracy (%)
	N	8408	105	52	128	9	96.62	99.21	99.86	98.21	97
	S	4	266	8	3	0	94.66	98.89	70.74	80.97	98.78
	V	2	5	699	14	0	97.08	99.26	90.90	93.89	99.11
	F	5	0	8	77	0	85.56	98.57	34.68	49.36	98.45
	Q	1	0	2	0	418	99.29	99.91	97.89	98.58	99.88

Table 7 Average of all tenfold ECG heartbeat Overall accuracy

Technique	Sampling Methods	Learning rate	Overall Accuracy (%)
CNN	Oversampling	1e-3	96.61
	Oversampling	1e-4	97.46
	SMOTE	1e-3	97.11
	SMOTE	1e-4	**98.23**

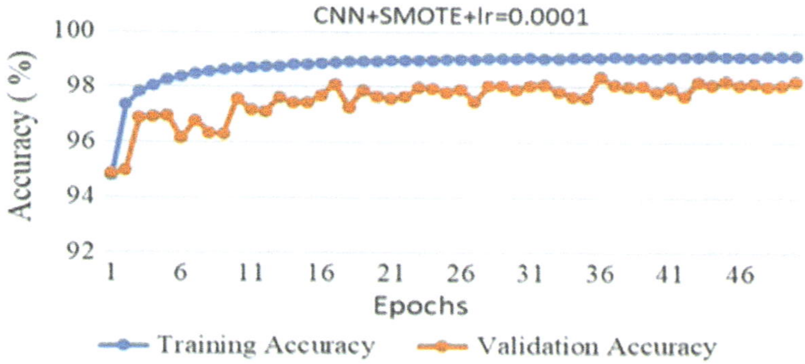

Fig. 5 The training and validation accuracy with epochs using SMOTE and learning rate 0.0001

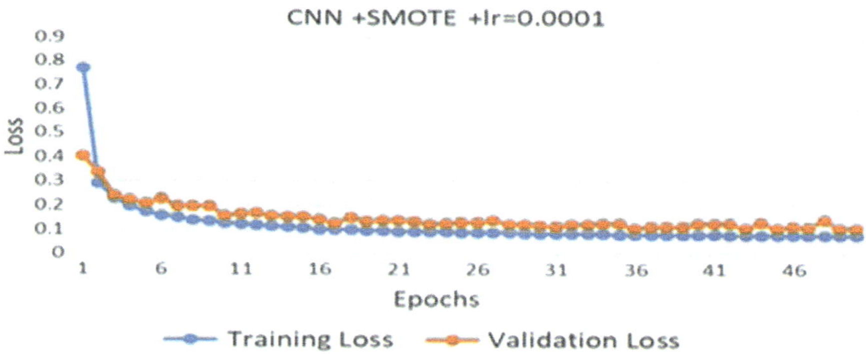

Fig. 6 The training and validation loss with epochs using SMOTE and learning rate 0.0001

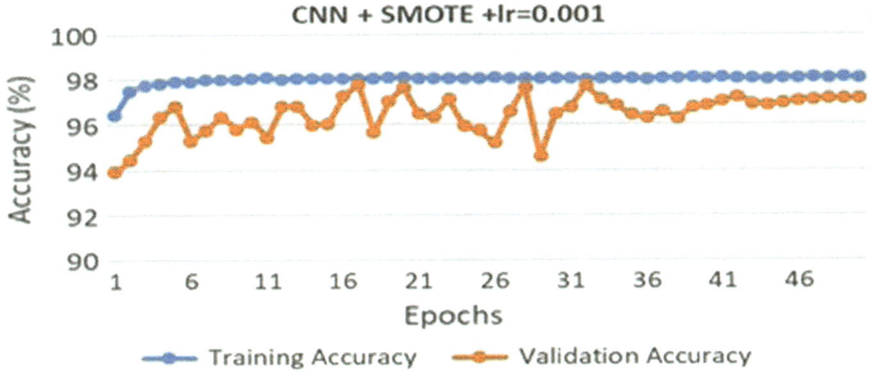

Fig. 7 The training and validation accuracy with epochs using SMOTE and learning rate 0.001

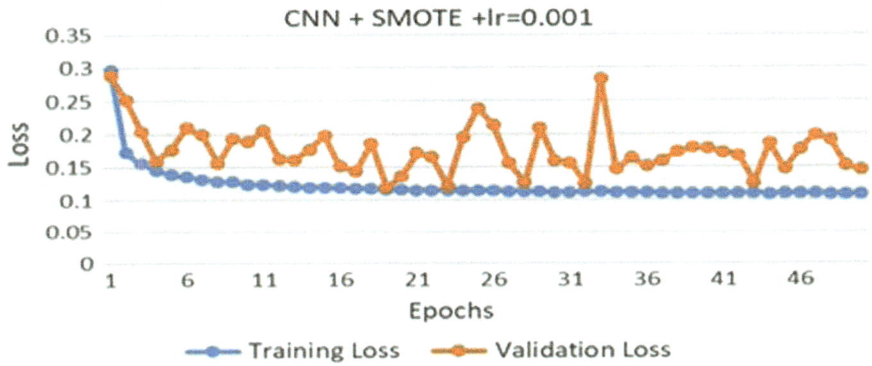

Fig. 8 The training and validation loss with epochs using SMOTE and learning rate 0.001

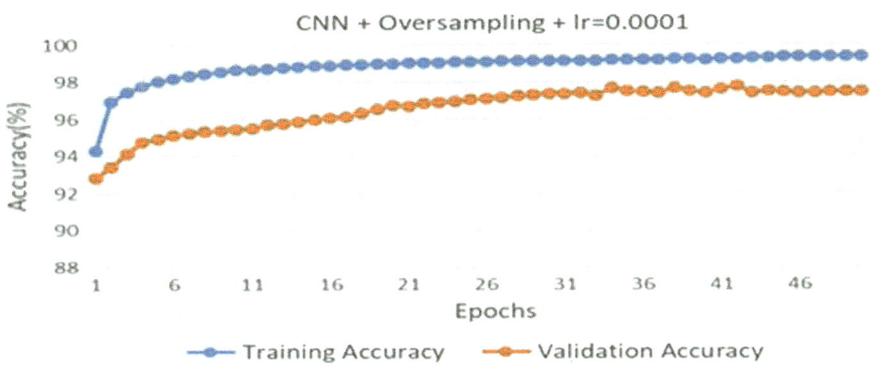

Fig. 9 The training and validation accuracy with epochs using oversampling and learning rate 0.0001

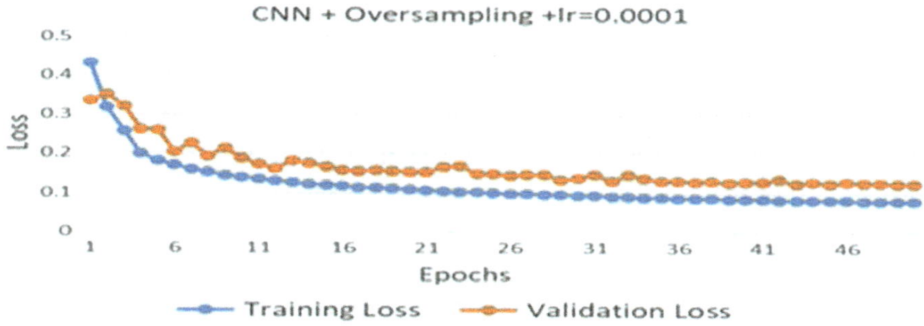

Fig. 10 The training and validation loss with epochs using oversampling and learning rate 0.0001

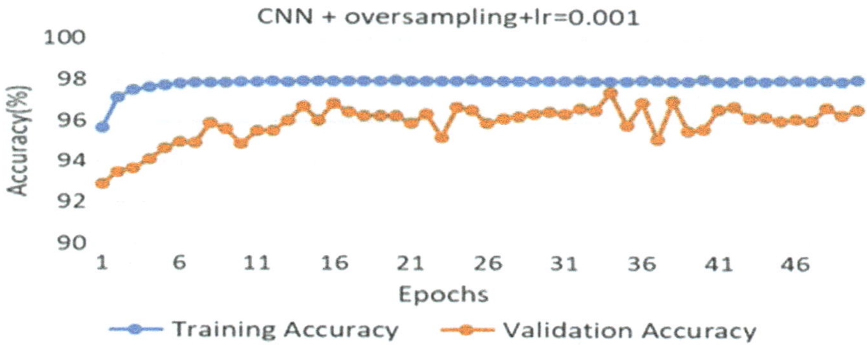

Fig. 11 The training and validation accuracy with epochs using oversampling and learning rate 0.001

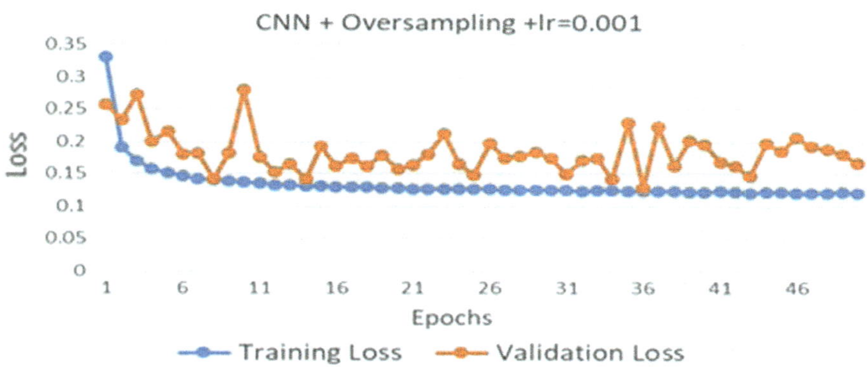

Fig. 12 The training and validation loss with epochs using oversampling and learning rate 0.001

Table 8 Performance comparison obtained from different ECG heartbeat classification approach on MIT-BIH arrhythmia database

Literature and year	Features	Classifier	Overall Accuracy (%)
Kutlu et al. [18] 2011	HOSF, MF, FFT	KNN	93.49
Martis et al. [50] 2013	PCA, HOS	SVM, Neural Network	93.48
Chazal et al. [51] 2013	R-R interval, MF	Linear discriminant	92.4
Tang et al. [52] 2014	Wavelet transform	RS-QNN	91.70
Lin et al. [53] 2014	R-R interval, MF	Linear discriminant	93.00
Zubair et al. [54] 2016	Raw data	Deep CNN model	92.70
Li, Taiyong et al. [55] 2016	WPE, R-R interval	Random forest, SVM, DT, PNN, K-NN	94.61
Shadmand et al. [56] 2016	PSO	Block based neural network	97.00
Acharya et al. [57] 2017	Raw data	Deep CNN model	94.03
Proposed work	Raw data	11-layer deep CNN model	98.23

5 Conclusion

In this research study, a completely automatic electrocardiogram signals classifier model has been developed. This model classified the ECG signal in five different classes according to the AAMI standards. A deep learning classification model was applied in the study trained and tested on MIT-BIH ECG database. It significantly improved the arrhythmia classification accuracy by overcoming the class imbalance problem. The class imbalance problem was managed by generating synthetic data using oversampling and SMOTE techniques. The simulation portrays CNN model to be an efficient system for ECG arrhythmia classification. The classification results have been presented using five measurements (Recall, specificity, precision, F1-score, Accuracy). We achieved 97.75% recall, 98.59% specificity, 92.32% precision, 94.96% F1-score, and 98.46% accuracy. These results are averaged across all ten-fold five heartbeat classes with the overall accuracy of CNN model being 98.23%. Our model outperforms the works mentioned in the paper [18, 50–57]. The developed CNN model can be applied on a CAD ECG system for the diagnosis of ECG heartbeats. This model can work as an assistive device for cardiologists by reading ECG heartbeat signals. Such types of devices have varying applications in polyclinics, the patients waiting times cardiologist workload can be reduced along with the optimization of the ECG signal processing cost.

References

1. Mehra R (2007) Global public health problem of sudden cardiac death. J Electrocardiol 40(6):S118–S122
2. World Health Organization (2017) Noncommunicable diseases: progress monitor 2017
3. Hadhoud MM, Eladawy MI, Farag A (2006) Computer aided diagnosis of cardiac arrhythmias. In: 2006 international conference on computer engineering and systems. IEEE
4. Singh S et al (2018) Classification of ECG arrhythmia using recurrent neural networks. Procedia Comput Sci 132:1290–1297
5. De Chazal P, O'Dwyer M, Reilly RB (2004) Automatic classification of heartbeats using ECG morphology and heartbeat interval features. IEEE Trans Biomed Eng 51(7):1196–1206
6. Alonso-Atienza F et al (2014) Detection of life-threatening arrhythmias using feature selection and support vector machines. IEEE Trans Biomed Eng 61(3):832–840
7. Khadra L, Al-Fahoum AS, Al-Nashash H (1997) Detection of life-threatening cardiac arrhythmias using the wavelet transformation. Med Biol Eng Comput 35(6):626–632
8. Sörnmo L, Laguna P (2005) Bioelectrical signal processing in cardiac and neurological applications, vol 8. Academic Press
9. Melillo P et al (2015) Wearable technology and ECG processing for fall risk assessment, prevention and detection. In: 2015 37th Annual International Conference of the IEEE engineering in medicine and biology society (EMBC). IEEE
10. Jain S et al (2017) QRS detection using adaptive filters: a comparative study. ISA Trans 66:362–375
11. Gacek A, Pedrycz E (eds) (2011) ECG signal processing, classification and interpretation: a comprehensive framework of computational intelligence. Springer Science & Business Media
12. Das MK, Ari S (2014) ECG beats classification using mixture of features.Int Sch Res Not 2014
13. Engin M (2004) ECG beat classification using neuro-fuzzy network. Pattern Recogn Lett 25(15):1715–1722
14. Lannoy G et al (2010) Feature relevance assessment in automatic inter-patient heart beat classification. Biosignals
15. Mar T et al (2011) Optimization of ECG classification by means of feature selection. IEEE Trans Biomed Eng 58(8):2168–2177
16. De Chazal P, Reilly RB (2006) A patient-adapting heartbeat classifier using ECG morphology and heartbeat interval features. IEEE Trans Biomed Eng 53(12):2535–2543
17. Ye C, Vijaya Kumar BVK, Coimbra MT (2012) Heartbeat classification using morphological and dynamic features of ECG signals. IEEE Trans Biomed Eng 59(10):2930–2941
18. Kutlu Y, Kuntalp D (2011) A multi-stage automatic arrhythmia recognition and classification system. Comput Biol Med 41(1):37–45
19. Llamedo M, Martínez JP (2011) Heartbeat classification using feature selection driven by database generalization criteria. IEEE Trans Biomed Eng 58(3):616–625
20. Das MK, Ghosh DK, Ari S (2013) Electrocardiogram (ECG) signal classification using s-transform, genetic algorithm and neural network. In: 2013 IEEE 1st international conference on condition assessment techniques in electrical systems (CATCON). IEEE
21. Shinde AA, Kanjalkar P (2011) The comparison of different transform based methods for ECG data compression. In: 2011 international conference on signal processing, communication, computing and networking technologies (ICSCCN). IEEE
22. Martis RJ, Acharya UR, Min LC (2013) ECG beat classification using PCA, LDA, ICA and discrete wavelet transform. Biomed Signal Process Control 8(5):437–448
23. Zhang L, Peng H, Yu C (2010) An approach for ECG classification based on wavelet feature extraction and decision tree. In: 2010 international conference on wireless communications and signal processing (WCSP). IEEE
24. Tang X, Shu L (2014) Classification of electrocardiogram signals with RS and quantum neural networks. International Journal of Multimedia and Ubiquitous Engineering 9(2):363–372

25. Rai HM, Trivedi A, Shukla S (2013) ECG signal processing for abnormalities detection using multi-resolution wavelet transform and Artificial Neural Network classifier. Measurement 46(9):3238–3246D
26. Alickovic E, Subasi A (2016) Medical decision support system for diagnosis of heart arrhythmia using DWT and random forests classifier. Journal of medical systems 40(4):108
27. Elhaj FA et al (2017) Hybrid classification of Bayesian and extreme learning machine for heartbeat classification of arrhythmia detection. In: 2017 6th ICT International student project conference (ICT-ISPC). IEEE
28. Khalaf AF, Owis MI, Yassine IDA (2015) A novel technique for cardiac arrhythmia classification using spectral correlation and support vector machines. Expert Systems with Applications 42(21):8361–8368
29. Rajesh KNVPS, Dhuli R (2018) Classification of imbalanced ECG beats using re-sampling techniques and AdaBoost ensemble classifier. Biomed Signal Process Control 41:242–254
30. Lu W, Hou H, Chu J (2018) Feature fusion for imbalanced ECG data analysis. Biomedical Signal Processing and Control 41:152–160
31. Roopa CK, Harish BS (2017) A survey on various machine learning approaches for ECG analysis. Int J Comput Appl 163(9)
32. Litjens G et al (2017) A survey on deep learning in medical image analysis. Med Image Anal 42:60–88
33. Acharya UR et al (2017) Deep convolutional neural network for the automated detection and diagnosis of seizure using EEG signals. Comput Biol Med
34. Wang Q et al (2017) A novel ensemble method for imbalanced data learning: bagging of extrapolation-SMOTE SVM. Comput Intell Neurosci
35. Goldberger AL et al (2000) PhysioBank, PhysioToolkit, and PhysioNet: components of a new research resource for complex physiologic signals. Circulation 101(23):e215–e220
36. Moody GB, Mark RG (2001) The impact of the MIT-BIH arrhythmia database. IEEE Engineering in Medicine and Biology Magazine 20(3):45–50
37. Luz EJDS et al (2016) ECG-based heartbeat classification for arrhythmia detection: a survey. Comput Methods Programs Biomed 127:144–164
38. Wikipedia contributors. "Feature scaling." Wikipedia, The Free Encyclopedia. Wikipedia, The Free Encyclopedia, 24 Jun. 2018. https://en.wikipedia.org/w/index.php?title=Feature_scaling&oldid=847274325
39. Chen S et al (2017) Heartbeat classification using projected and dynamic features of ECG signal. Biomed Signal Process Control 31:165–173
40. Chen Y-S (2016) An empirical study of a hybrid imbalanced-class DT-RST classification procedure to elucidate therapeutic effects in uremia patients. Medical & biological engineering & computing 54(6):983–1001
41. Radivojac P et al (2004) Classification and knowledge discovery in protein databases. J Biomed Inform 37(4):224–239
42. Zheng Z, Xiaoyun W, Srihari R (2004) Feature selection for text categorization on imbalanced data. ACM Sigkdd Explorations Newsletter 6(1):80–89
43. Prati RC, Batista GEAPA, Monard MC (2008) A study with class imbalance and random sampling for a decision tree learning system. In: IFIP international conference on artificial intelligence in theory and practice. Springer, Boston, MA
44. Chawla NV et al (2002) SMOTE: synthetic minority over-sampling technique. J Artif Intell Res 16:321–357
45. Karpathy A (2016) Cs231n convolutional neural networks for visual recognition. Neural networks 1
46. Zebik M et al (2017) Convolutional neural networks for time series classification. In: International conference on artificial intelligence and soft computing. Springer, Cham
47. Zheng Y et al (2014) Time series classification using multi-channels deep convolutional neural networks. In: International conference on web-age information management. Springer, Cham
48. Krizhevsky A, Sutskever I, Hinton GE (2012) Imagenet classification with deep convolutional neural networks. In: Advances in neural information processing systems

49. Bani-Hasan, AM, El-Hefnawi MF, Kadah MY (2011) Model-based parameter estimation applied on electrocardiogram signal. J Comput Biol Bioinform Res 3(2):25–28
50. Martis RJ et al (2013) Cardiac decision making using higher order spectra. Biomed Signal Process Control 8(2):193–203
51. De Chazal P (2013) A switching feature extraction system for ECG heartbeat classification. In: 2013 computing in cardiology conference (CinC). IEEE
52. Tang X, Shu L (2014) Classification of electrocardiogram signals with RS and quantum neural networks. Int J Multimed Ubiquitous Eng 9(2):363–372
53. Lin C-C, Yang C-M (2014) Heartbeat classification using normalized RR intervals and morphological features. Math Probl Eng
54. Zubair M, Kim J, Yoon C (2016) An automated ECG beat classification system using convolutional neural networks. In: 2016 6th international conference on IT convergence and security (ICITCS). IEEE
55. Li T, Zhou M (2016) ECG classification using wavelet packet entropy and random forests. Entropy 18(8):285
56. Shadmand S, Mashoufi B (2016) A new personalized ECG signal classification algorithm using block-based neural network and particle swarm optimization. Biomed Signal Process Control 25:12–23
57. Acharya UR et al (2017) A deep convolutional neural network model to classify heartbeats. Comput Biol Med 89:389–396

Chapter 3
Breast Cancer Identification and Diagnosis Techniques

V. Anji Reddy and Badal Soni

Abstract Identification of disease in humans accurately is very difficult and also important for further treatment. One of the major tasks for a doctor is the identification of the disease. Once the disease is identified then it is very easy to perform diagnosis for the patient. In this chapter, we reviewed and presented various machine learning and deep learning algorithms for disease identification. Mainly we are presenting on one of the most occurring diseases in women, that is breast cancer. Here we are presenting the various methodology and algorithms for the identification of breast cancer. Some of the methodologies and algorithms are Support Vector Machine (SVM), Biclustering mining and Adaboost Algorithm, Convolution Neural Network (CNN), Recurrent Neural Network (RNN), Breast Imaging reporting and data system (BI-RADS), ICD-9 diagnosis codes from an existing EHR data repository, Hierarchical Attention bidirectional networks (HA-BiRNN), Clinical decision support system, Outlier Detection Algorithm (ODA).

Keywords Breast cancer · SVM · CNN · RNN · CLAHE · Medical image

1 Introduction

Breast cancer is said to be the foremost cause of fatality in women. As the disease is anonymous, premature recognition and analysis is the solution for controlling breast cancer, which directly reduces the death rate of lives at a minimal cost. Taking usual screening tests is the most consistent way to discover breast cancer in the beginning stage. For screening breast cancer, mammography has emerged as the gold standard, which has the capability of detecting cancer at the earlier stage. Moreover, it is capable of overcoming the clinical difficulties such as architectural distortion, asymmetries among the breasts and calcifications masses related to benign fibrosis.

V. Anji Reddy (✉) · B. Soni
National Institute of Technology, Silchar, India
e-mail: anjisoftware@hotmail.com

B. Soni
e-mail: soni.badal88@gmail.com

J. K. Rout et al. (eds.), *Machine Learning for Intelligent Decision Science*,
Algorithms for Intelligent Systems, https://doi.org/10.1007/978-981-15-3689-2_3

The abnormalities of mammographic are classified into two most important groups: masses and calcifications.

As per the South Australian survey, 41% of the death rates have been reduced by the mammography screening model. During the screening, several investigations have portrayed that an approximate of 11–25% of breast cancers were not detected at the premature stage. The homogeneous regions are uniformly divided into significant Region Of Interest (ROI) using the mammogram image segmentation. The segmentation techniques are categorized into two diverse groups depending on the regions, which are to be segmented, like breast region segmentation and ROI segmentation. The region-oriented segmentation is the action of segmenting the uncertain regions for abnormalities that are to be examined. The second one is the process of partitioning the mammogram image into a background and breast region, by which the background issues could be resolved.

Many of the approaches used for processing mammograms were dependent on features extraction and classification. WT offers better multi-direction and multi-resolution aspects and hence, it is exploited mostly during the processing of mammograms. In addition, techniques like Principle Component Analysis (PCA) could be exploited for dimension reduction purposes and also for converting the descriptors to a compact form. Contourlet transform and top hat are certain techniques that help in improving the characteristics of micro-calcifications and it aids in eliminating the image noise.

1.1 Clinical Decision Support Systems

CDS [1] frameworks offer patients, clinicians, staff, and other persons with person-specific and knowledgeable data, sharply filtered and offered at proper times to improve health and medical care. The US institution, "Institute of Medicine" has documented the issues with the quality of health care. For previous years, it has supported health IT to increase the features of diagnosis. From 2004, while the Federal Government encouraged the significance of EMRs, there was a gradual increase in the adoption of health IT [2]. Such IT appliances are a way to develop the value of health care.

The widespread usage of CDS is for representing clinical requirements like guarantying precise diagnosis, viewing in an appropriate manner for avoidable diseases, or prevention of undesirable drug measures. On the other hand, CDS could also potentially reduce costs, enhance efficiency, and minimize patient problems as well. Actually, CDS could deal with all areas concurrently—for instance, by making healthcare professionals to be aware of redundant diagnosis. Moreover, the CDS aims to assist the clinician in reducing the burden by indicating the respective errors occurring during the diagnosis. The CDS might suggest recommendations; however the clinician should sort out the data, reconsider the suggestions, and make a choice regarding the treatment. Table 1 demonstrates the objectives of CDS as per [3].

Table 1 Objectives of CDS

Objectives of CDS	Intention of users	Key issues
Offer data when a user is uncertain what to do	High	Ease and speed of access
Reminder of activities user aims to do, but shouldn't have to keep in mind	High	Timing
Rectify errors of users or suggest user change plans	Low	Regular: timing, autonomy and user control; On demand: autonomy, simpler assessing, speediness and response for user control

2 Imaging Techniques

Around one in eight women is suffering from breast cancer in America, and, of these, approximately 30% will die due to the disease in the end. Therefore premature detection and precise analysis of breast cancer are significant. The conventional clinical techniques were exploited for breast cancer diagnosis and screening, however, they include several negative aspects. X-ray mammography, for instance, has an approximate of 22% FNR in women under the age of fifty and at times, it cannot exactly differentiate the malignant and benign tumours. Although the FPR of mammography is below 10%, 18% of ladies without cancer have gone through a biopsy after ten mammograms.

Methods like MRI and US are sometimes exploited along with X-ray mammography; however, they have certain shortcomings like low sensitivity (US), low throughput, high cost and limited specificity (MRI). Therefore, novel techniques are required for detecting cancers failed to spot by mammography and thereby, the FPR is minimized while monitoring tumour development during cancer treatment. NIR, DOT and spectroscopy are equipment, which relies on contrast and as a result, they include the ability to develop specificity and sensitivity of breast cancer detection. Till now, different tumours (i.e. above 30 tumours) have been reported by numerous investigation groups. These results usually point out that optical techniques are proficient of distinguishing malignant and benign tumours from healthier tissue. Manifestation of difference among malignant and benign tumours thus remains insufficient, either due to the imperfect three-dimensional imaging model or due to the shortage of benign lesion data. Figure 1 shows the examples of breast changes that may be seen on a mammogram.

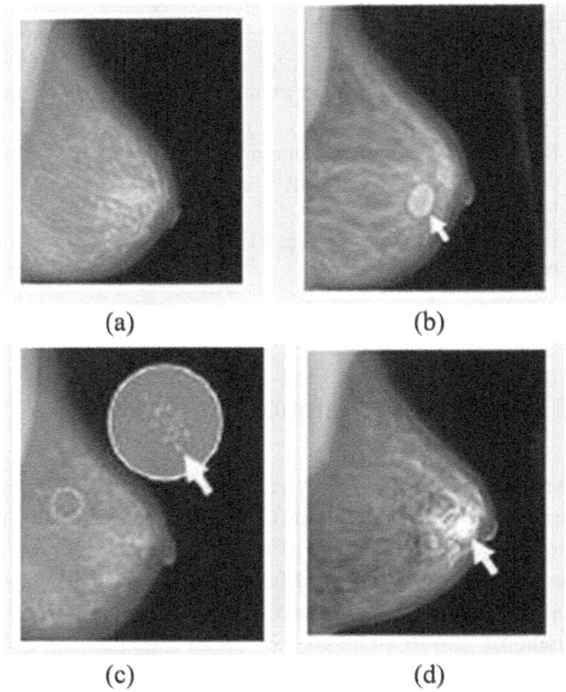

Fig. 1 Pictorial illustration of **a** normal mammogram **b** benign cyst (not cancer) **c** breast calcifications **d** malignant breast cancer

3 Pre-processing Techniques

The major objective of the pre-processing technique is to enhance the quality of the image by reducing or removing the disparate and additional portions in the backdrop of mammogram images. "Mammograms are medical images that are complex to understand". Therefore pre-processing is necessary for enhancing the quality. In this stage, the mammogram undergoes segmenting and feature extracting processes. The high-frequency components and noise are eliminated by filters. The pre-processing techniques are categorized into contrast adjustment and intensity based schemes, Filtering-based models, Binarization-based models and morphological operation based models. The contrast adjustment can be enhanced using the techniques such as CLAHE technique, HM-LCE technique, Linear Starching, Convolution Mask Enhancement, Enhancement by Point Processing and so on. The filtering-based models include techniques such as temporal filter, Mean filter, Median filtering, AMF, Han filter, Wiener filter and spatial filter. The binarization-based models can be categorized into local binarization and global binarization models. Moreover, the morphological operation based models include structuring element, dilation, erosion and opening and closing process. Figure 2 shows the taxonomy of the techniques adopted for pre-processing. From the above-mentioned techniques, certain techniques are elaborated in the following sections.

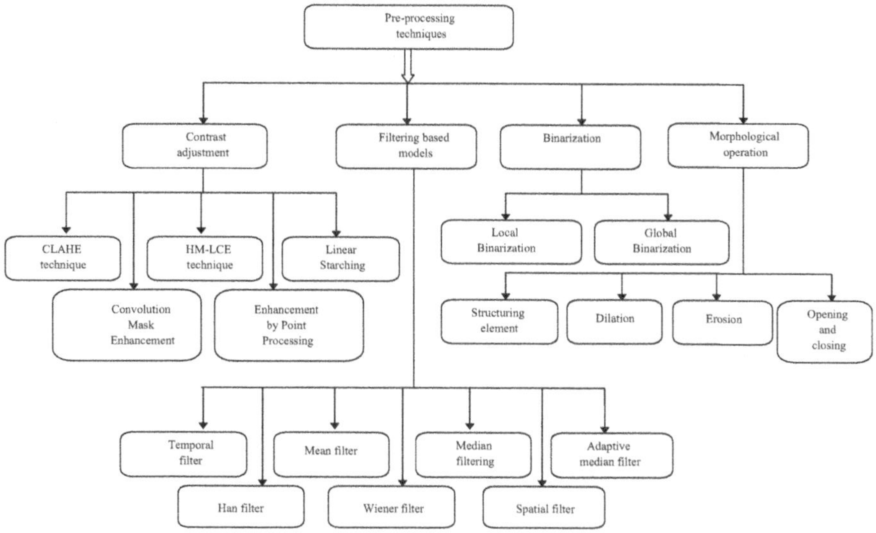

Fig. 2 Taxonomy of the techniques adopted for pre-processing

3.1 *Mean Filter*

The objective of the mean filters is to enhance the quality of image. Here, the filter is applied for all the pixels with the standard value of neighbourhood intensities. As it locally minimized the variance, it is simpler to be performed. The drawbacks of average filter are as follows.

(a) Averaging operations causes the blurring of the image, which in turn affects the localization of features.
(b) If the averaging functions deployed to an image that is damaged by noise, it gets diffused and attenuated; however it could not be eliminated.
(c) A particular pixel with a much unreliable value affects the average value of entire pixels in its neighbourhood.

3.2 *Median Filtering*

"A median filter is a nonlinear filter that is efficient in removing salt and pepper noise and median tends to keep the sharpness of image edges during noise removal". The components of the median filter are (a) "Max-median filter (b) weighted median filter (c) Centre-weighted median filter". The noise could be eliminated efficiently when there is increase in window size.

3.3 AMF Technique

AMF operates on a rectangle area S_{xy}. It alters the dimension of S_{xy} throughout the filter process based on specific circumstances as mentioned here. Every output pixel includes the median value in the "3-by-3 neighbourhood" surrounding the related pixel in input images. Here, the image edges are substituted by 0's. The filter output is a specific value that replaced the present pixel value at (x, y) point.

$$Z_{\max} = \text{utmost value of pixel in } S_{xy}$$
$$Z_{\min} = \text{least value of pixel in } S_{xy}$$
$$Z_{xy} = \text{value of pixel at coordinates } (x, y)$$
$$S_{\max} = \text{utmost permitted the size of } S_{xy}$$
$$Z_{\text{med}} = \text{median value of pixel in } S_{xy}$$

Adaptive Median filtering is exploited for smoothening the noise from 2D waveforms devoid of preserved images and edge blurring. Thus, it is appropriate for improving the mammogram images. This filtering method is exploited in mammogram for label, orientation, artefact elimination, segmentations and improvement. The pre-processing is also concerned in making masks for pixels with maximum intensity for minimizing the resolutions and for segmenting the breast.

3.4 Wiener Filter

This filter attempts to construct a finest estimation of the real image by implementing a MMSE parameter among the original and estimated image. "The wiener filter is an optimal filter and it intends to reduce the MSE present in the image and it has the ability to handle both the deterioration function along with the noise". From the deterioration model, the error among the estimated signal $f(a, b)$ and the input signal $f(a, b)$ is specified by Eq. (1). The error is given by Eq. (2), and the MSE is specified by Eq. (3).

$$E(A, B) = F(A, B) - F(A, B) \tag{1}$$

$$[F(A, B) - F(A, B)]^2 \tag{2}$$

$$E\{[F(A, B) - F(A, B)]^2\} \tag{3}$$

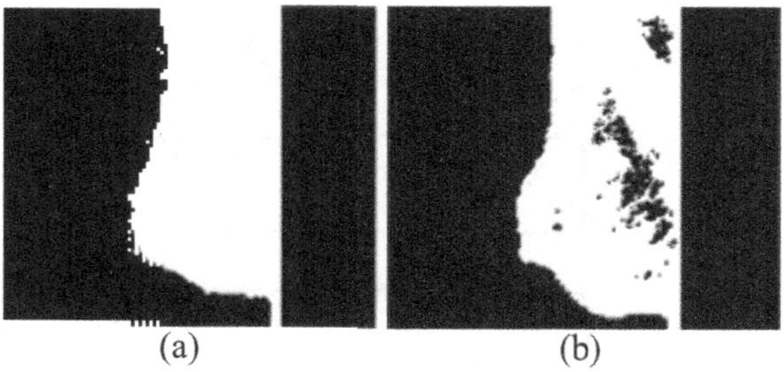

Fig. 3 Pictorial illustration of **a** original image **b** enhanced image

3.5 *CLAHE Technique*

This model is an updated version of AHE, which is a computer-based image processing model that enhances the image contrast. Usually, AHE operates on only a small portion of the image and not in the entire image. The histogram is evaluated for every portion that helps for redistributing the lightness image values. Thus, this model could increase the local contrast and it also seeks to minimize the noise factors. Figure 3 shows the model of original and enhanced breast image using CLAHE method [4].

3.6 *HM-LCE Technique*

HM-LCE technique has two-level processing that includes LCE and HM methods. Figure 4 demonstrates the steps concerned in the HM-LCE scheme. The ability of this contrast improvement technique offers contrast improvement in image with respect to both objective as well as subjective quality when distinguished over improvement approaches [5].

4 Feature Extraction Techniques

Each pixel in a digital mammogram is addressed by a feature vector that is evaluated based on their adjacent pixels. For a pixel, the neighbourhood is represented by a squared window W_n of $n \times n$ pixels. The window W_n is pre-processed by the normalization of gray-level values of its pixels, such that the standard deviation and mean is one and zero respectively. Subsequently, this window is processed in diverse models based on diverse feature extraction techniques like Sobel Filter, SGLDM, gray-map, SFUM and AFUM. These techniques (except AFUM) attain the feature vectors, where the dimensionality is based on the value of the nth constraint. Moreover, the dimensions of the feature vectors are minimized by deploying the PCA. Rather than producing the feature vectors, the AFUM model offers a particular feature for every

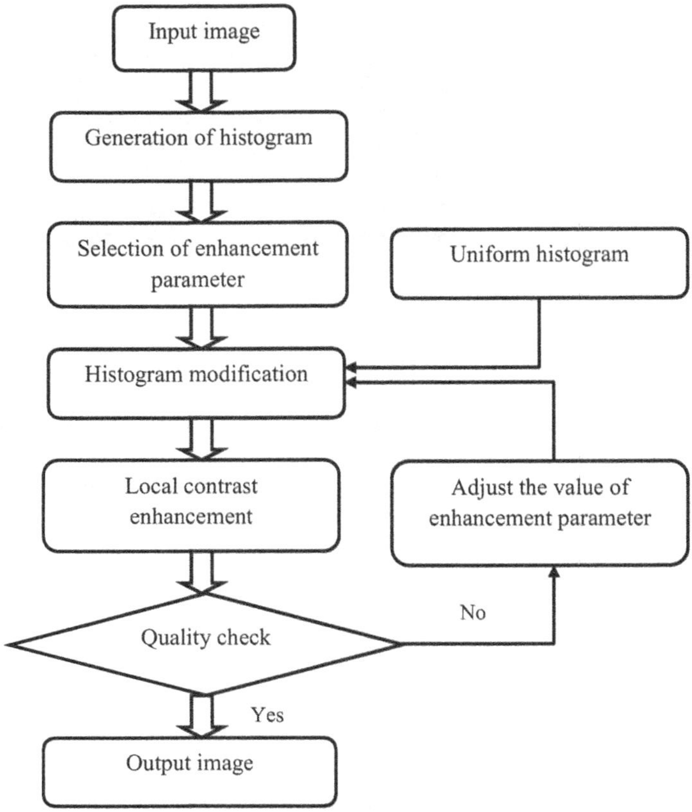

Fig. 4 Flow chart representation of HM-LCE model

pixel that is exploited as a confidence value for classifying it. The taxonomy of the types of feature extraction methods is given by Fig. 5.

4.1 Gray-Map

Gray-map offers very reliable outcomes when evaluated over other more complicated feature extraction techniques that are generally used in image detection. To assess this scheme, n^2 vectors are created with the standardized values of gray-level in W_n window. After that, these vectors are dimensionally reduced using the PCA model and accumulated as samples.

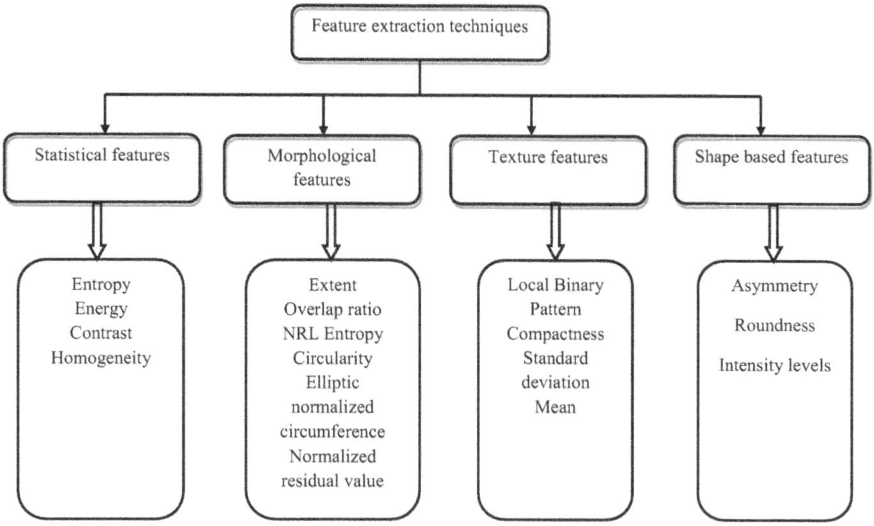

Fig. 5 Taxonomy of the types of feature extraction methods

4.2 Sobel

It is a general method to improve the edges of objects in an image. It comprises of 2 kernels that identify vertical and horizontal transformations in an image. If these kernels are exploited in an image, the outcomes could be deployed to evaluate the direction and edges magnitude of the image. Sobel approach also requires reduced computation time and it is perfectly adjusted to identify the edges in three pixels. However, it is strongly susceptible to noise and thus generates incoherent contours.

4.3 SGLDM

More segmentation techniques in medical images are dependent on "texture descriptors" like Co-occurrence Matrices (also well-known as SGLDM, Fractal features, and other types of textural features.

4.4 AFUM

Heath and Bowyer describe a mass recognition approach known as AFUM. Assume a pixel p_{ij} at a distance r_2 from p_{ij} with minimal intensity over the entire other pixels at centre (i, j) and radius r_1, is evaluated. This computation is made for both r_1 and r_2, and the mean of all pixel values is evaluated.

4.5 SFUM

Even though it is better to include a particular feature for every pixel, there may be chances of occurrence of data loss while the mean of features sets is evaluated. For avoiding it, a modification of the AFUM algorithm is proposed. Rather than evaluating the average, the minimum values of SFUM is taken into consideration, which offers a feature vector as an alternative to single scalar. If the vector dimensions are very higher, it can be minimized using PCA.

5 Machine Learning Approaches

Certain techniques that were generally adopted for classifying the breast cancer images are described in this section.

5.1 Support Vector Machine

This technique is an amalgamation of RFE and SVM [6]. RFE is a technique, which operates by choosing dataset features depending on the least feature value in a recursive manner. Accordingly, SVM-RFE is operated by removing the inappropriate features (lowest weight feature) in all iterations. For this cause, it is required to compute the weights of features in ascending order. Generally, this technique is separated into three phases:

Phase 1: The database is trained by means of SVM to compute the entire weight of features.

To determine the weight of features, SVM-train can be exploited that generates a α classifier. The weight function is portrayed by Eq. (4).

$$w = \sum_{1}^{k} \alpha_k y_k x_k \tag{4}$$

In the above equation, x_k refers to the training data from k-th feature and y_k points out the class label of k-th feature

Phase 2: Compute the ranking measure

To categorize the features depending on weight, "criterion ranking" is required, whose task is portrayed by Eq. (5).

$$c_k = w_k^2, \quad k = 1, 2, \ldots, |S| \tag{5}$$

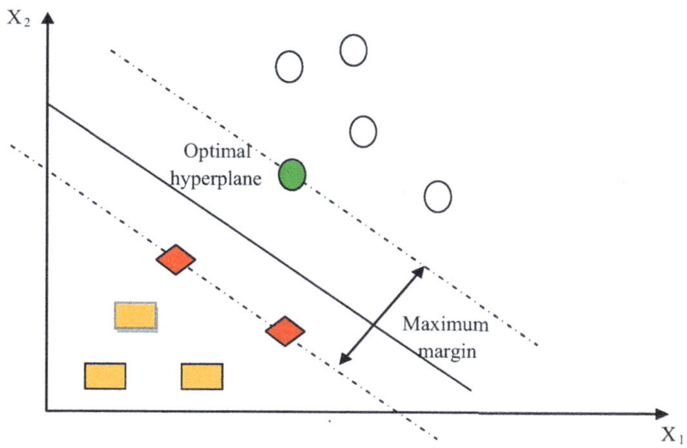

Fig. 6 General model of SVM

Phase 3: Sort the features on the basis of weight value and remove the feature with the least value of weight in all iterations. The basic model of the SVM framework is given by Fig. 6.

As per [6], the comparison of cancer data is attained as shown by Table 2.

5.2 Biclustering and Adaboost Techniques

The biclustering model encompasses the below three phases [7].

Phase 1: Deploy "hierarchical clustering" technique with a threshold distance, *Thc* on every column to split to numerous clusters. These entire clusters are recognized as BS.

Phase 2: Carry out a heuristic exploration for biclusters with reduced MSRS depending on this BS.

(a) For every BS, enlarge the columns to create an initial sub matrix *N*.
(b) The columns and rows of *N* are traversed and the column/row are removed if the novel sub matrix has the least MSRS in every process.
(c) Replicate the prior process till the MSRS of the novel sub matrix is minimal than predetermined *Thc*, and subsequently, a suitable bicluster is attained.

Table 2 Comparison of cancer data with 20% testing data

Type of cancer	Number of features (%)	Naïve Bayes classifier				SVM classifier			
		Accuracy	Precision	Recall	Time	Accuracy	Precision	Recall	Time
Prostate cancer	25	90	100	88.23	0.0036	90	100	88.23	0.0047
	50	90	100	88.23	0.0040	85	100	82.35	0.0034
	75	90	100	88.23	0.0034	85	100	82.35	0.0038
	100	85	100	82.35	0.0037	85	100	82.35	0.0035
Breast cancer	25	95.61	95.65	93.61	0.0103	95.61	95.65	93.61	0.0088
	50	94.73	95.65	91.48	0.0114	95.61	95.65	93.61	0.0100
	75	95.61	95.65	93.61	0.0110	95.61	95.65	93.61	0.0110
	100	95.61	95.65	93.61	0.0189	95.61	95.65	93.61	0.0116

Phase 3: Eliminate redundant biclusters that are completely overlapped by outsized ones and also remove certain repeated biclusters. After the mining of biclusters, the discovered bicluster patterns are transformed into analytic rules.

AdaBoost:

AdaBoost is a most renowned ensemble technique and it is proficient of enhancing the accurateness of classification by combining several weak classifiers. The bicluster-oriented classifiers can also be integrated with a strong ensemble classifier for superior generalization performance. During training, diverse weights are allocated and decisions are made depending on "weighted majority voting". The weight distribution at all iterations is updated based on the classification outcome of every bicluster-oriented classifier. The weights of perfectly-classified cases turn out to be lower, while the weights of those misclassified cases turn out to be higher. Figure 7 shows the modelling of adaboost algorithm. The comparison results among different CAD systems using Biclustering + AdaBoost model in [7] is given by Table 3.

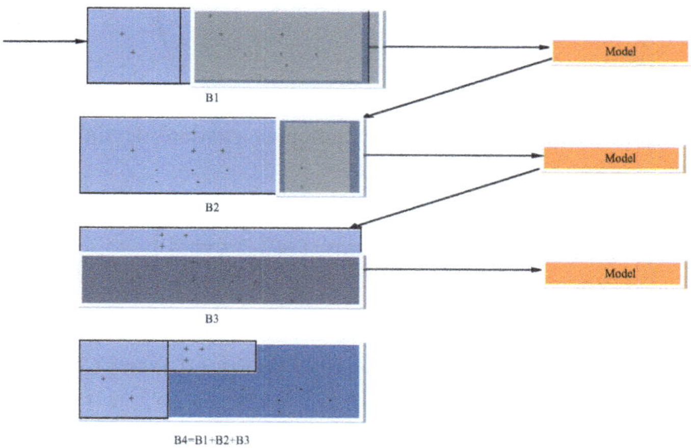

Fig. 7 General model of adaboost algorithm

Table 3 Comparison of diverse CAD systems

Classifier	Accuracy (%)	Specificity (%)	Sensitivity (%)
Fuzzy cerebellar mode NN	92.31	91.18	93.55
Fuzzy SVM	94.25	96.08	91.67
SVM	94.4	94.4	94.3
Biclustering + AdaBoost	95.75	95.12	96.26

5.3 CNN Classifier

CNN exploits the spatial data [8] amongst the image pixels and therefore, they depend on "discrete convolution". Accordingly, a gray scale image is presumed to be portrayed by Eq. (6), with a size of $a_1 \times a_2$.

$$\mathrm{Im} : \{1, \ldots, a_1\} \times \{1, \ldots, a_2\} \to P \subseteq \Re, (i, j) \mapsto \mathrm{Im}_{i,j} \tag{6}$$

Assume filter $D \in \Re^{2ge_1+1\times 2ge_2+1}$ and for image Im, the "discrete convolution with filter D" is specified by Eq. (7) and filter D is computed as per Eq. (8).

$$(\mathrm{Im} * D)_{l,r} := \sum_{v=-ge_1}^{ge_1} \sum_{u=-ge_2}^{ge_2} D_{v,u} \mathrm{Im}_{l+v,r+u} \tag{7}$$

$$D = \begin{pmatrix} D_{-ge_1,-ge_2} & \cdots & D_{-ge_1,-ge_2} \\ \vdots & D_{0,0} & \vdots \\ D_{ge_1,-ge_2} & \cdots & D_{ge_1,ge_2} \end{pmatrix} \tag{8}$$

The generally used filter for the purpose of smoothing is the discrete Gaussian filter $D_{H(\sigma)}$ as shown by Eq. (17), in which σ indicates standard deviation of "Gaussian distribution".

$$\left(D_{H(\sigma)}\right)_{l,r} = \frac{1}{\sqrt{2\pi\sigma^2}} \exp\left(\frac{l^2 + r^2}{2\sigma^2}\right) \tag{9}$$

The output of layer s includes $n_1^{(S)}$ feature maps with $n_2^{(S)} \times n_3^{(S)}$ dimension. The ith feature map indicated by $Xe_i^{(S)}$ is specified by Eq. (10), and $C_i^{(S)}$ indicates bias matrix and $D_{i,j}^{(S)}$ signifies the filter size of $2ge_1^{(S)} + 1 \times 2ge_2^{(S)} + 1$ that links the ith feature map in S layer with jth feature map in layer $(S - 1)$.

$$Xe_i^{(S)} = C_i^{(S)} + \sum_{j=1}^{n_1^{s-1}} D_{i,j}^{(S)} * Xe_j^{(S-1)} \tag{10}$$

In addition, the border effects influences $n_2^{(S)}$ and $n_3^{(S)}$ and the output feature maps have a size as revealed by Eq. (11).

$$n_2^{(S)} = n_2^{(S-1)} - 2ge_1^{(S)} \text{ and } n_3^{(S)} = n_3^{(S-1)} - 2ge_2^{(S)} \tag{11}$$

On considering the convolution layer as given in Eqs. (20) and (12) can be remodelled. Every $X_i^{(S)}$ in layer S includes $n_2^{(S)}.n_3^{(S)}$ units in 2D matrix. Equation (13) shows the output by considering the unit at position (l, r).

$$\left(Xe_i^{(S)}\right)_{l,r} = \left(C_i^{(S)}\right)_{l,r} + \sum_{j=1}^{n_1^{(S-1)}} \left(D_{i,j}^{(S)} * Xe_j^{(S-1)}\right)_{l,r} \tag{12}$$

$$= \left(C_i^{(S)}\right)_{l,r} + \sum_{j=1}^{n_1^{(S-1)}} \sum_{v=-ge_1^S}^{ge_1^S} \sum_{u=-ge_2^S}^{ge_2^S} \left(L_{i,j}^{(S)}\right)_{v,u} \left(D_{i,j}^{(S)}\right)_{l,r} \left(Xe_j^{(S-1)}\right)_{l+v,r+u} \tag{13}$$

The input S is specified by $n_1^{(S)}$ feature maps and its output again includes $n_1^{(S)} = n_1^{(S-1)}$ feature maps if S is a nonlinearity layer as denoted by Eq. (14), in which f denotes the activation function.

$$Xe_i^{(S)} = f\left(Xe_i^{(S-1)}\right) \tag{14}$$

If S and $S-1$ is a fully connected layer, Eq. (15) can be applied, in which $y^{(S-1)}$, $we^{(S)}$ and $z^{(S)}$ indicates the respective vectors and matrix representations of outputs f the weights, $we_{i,k}^{(S)}$ and actual inputs $z_i^{(S)}$ correspondingly. Or else, layer S considers $n_1^{(S-1)}$ feature maps of size $n_2^{(S-1)} \times n_3^{(S-1)}$ as input and ith input in layer S evaluates Eq. (15). Here, $we_{i,j,l,r}^{(S)}$ denotes the weights linking the unit at (l, r) position in layer $(S-1)$ and ith unit in layer S. Figure 8 shows the general model of CNN framework [9].

$$z_i^{(S)} = \sum_{k=0}^{n^{(S-1)}} we_{i,k}^{(S)} y_k^{S-1} \quad \text{or} \quad z^{(S)} = we^{(S)} y^{(S-1)} \tag{15}$$

$$y_i^{(S)} = f\left(z_i^{(S)}\right) \text{ with } z_i^{(S)} = \sum_{j=1}^{n_1^{(S-1)}} \sum_{l=1}^{n_2^{(S-1)}} \sum_{r=1}^{n_3^{(S-1)}} we_{i,j,l,r}^{(S)} \left(X_k^{S-1}\right)_{l,r} \tag{16}$$

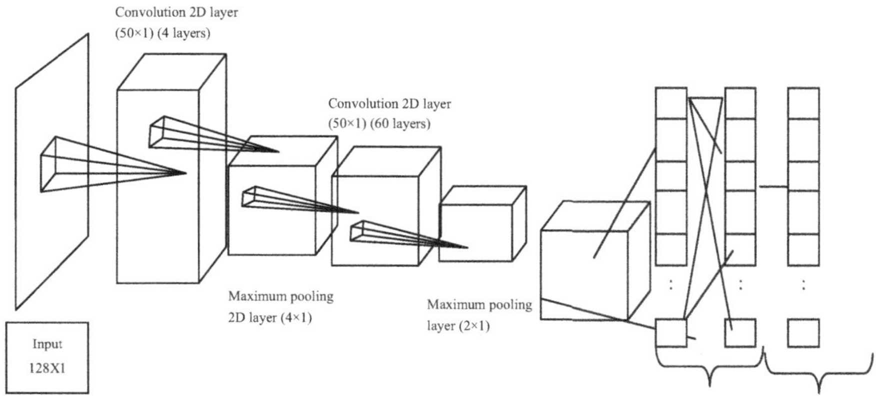

Fig. 8 General model of CNN framework

Table 4 Analysis on CNNI-BCC

Methods	Accuracy	Sensitivity	False negative rate	Specificity	False positive rate
Neuroph MLP	35.75	97.37	2.63	81.25	77.05
BPNN	77.08	75.00	–	81.25	–
Adaboost	81.25	78.12	–	87.5	–
CNNI-BCC	90.50	89.47	10.51	90.71	9.29

The general model of CNN framework is given by Fig. 8 and accordingly, Table 4 shows the analysis outcomes attained using CNNI-BCC [8].

5.4 RCNN

RNNs [10] are the group of NN, which are "deep" in sequential dimension and were exploited widely in time-sequence modelling. In contrast to a traditional NN, RNNs are capable of processing the data points where the activation at every step is based on prior step. Usually, for a specified sequence data $x = (x_1, \ldots, x_T)$, an RNN update its recurring hidden state h_t as per Eq. (17), in which h_t and x_t denotes the recurrent hidden state and the data values at time step t, correspondingly, and $\varphi(.)$ refers to "nonlinear activation function", namely a hyperbolic or sigmoid tangent.

$$h_t = \begin{cases} 0, & \text{if } t = 0 \\ \varphi(h_{t-1}, x_t) & \text{otherwise} \end{cases} \tag{17}$$

In addition, the RNN might have an output $y = (y_1, \ldots, y_T)$. In the conventional RNN approach, the update equation of the recurrent hidden state in Eq. (17) is formulated as per Eq. (18).

$$h_t = \varphi(Wx_t + Uh_{t-1}) \tag{18}$$

In the above equation, U and W denotes the recurrent hidden units activation at the preceding step and input coefficient matrices at the current step, correspondingly. Table 5 shows the comparison of accuracy attained using [10].

5.5 BI-RADS

The mammographic BI-RADS [11] lexicon introduced by the ACR has facilitated more reliable management and evaluation of non-palpable lesions and it further

Table 5 Comparison of accuracy

Method	Patch-wise accuracy (%)	Image-wise accuracy (%)
CNN	66.7	77.8
Densely connected CNN	–	83.0
CNN with image deformation	–	83.0
Deep CNN	–	87.2
Ensemble classifier	79.0	87.5
RCNN based model	82.1	91.3

assists in the classification of lesions depending on the likelihood of malignant cells. The BI-RADS dictionary was extended to take account of breast US in 2003.

The BI-RADS was introduced to normalize mammographic reporting. Conditions have been determined for portraying the lesion features, breast density, recommendations and impression. The technique has attained enhanced results in mammographic treatment while considering observers' scaled values, patient age and morphologic features. These descriptors were weighted for analytical power, and the output to the observer was a recommended approximation of the chance of malignancy. Likewise, the input of BI-RADS descriptors into an ANN was revealed to enhance the PPV of breast biopsy. The last condition of the "Mammography Quality Standards Act" makes use of the BI-RADS final evaluation categories on all mammographic analysis. Inconsistency in the appliance of the BI-RADS terms has not been extensively analysed in practice. The results for positive to negative findings, PPV and characteristics of BI-RADS 3–5 classifications as per [12] are demonstrated by Tables 6, 7 and 8 respectively.

Table 6 Analysis on BI-RADS category

	BI-RADS 4 (suspicious)	BI-RADS 3 (probably benign)	BI-RADS 5 (highly suspicious)
Negative *n* (%)	94 (52.2)	547 (97.3)	3 (3.2)
Positive *n* (%)	86 (47.8)	15 (2.7)	92 (96.8)
Total *n* (%)	180 (100)	562 (100)	95 (100)

Table 7 PPV related to BI-RADS 3–5

BI-RADS	5	4	3
PPV	0.97	0.48	0.03

Table 8 Characteristics of classifications on BI-RADS 3–5

BI-RADS 3–5	
Accuracy	0.87
Specificity	0.85
Sensitivity	0.92

Table 9 Analysis on the HBRNN schemes

Methods	Accuracy	Recall	F1-score
$HABRNN_g$	0.8429	0.8289	0.7814
$HABRNN_c$	0.8597	0.8471	0.7978
HBRNN	0.8250	0.8009	0.7159
$HARNN_l$	0.8631	0.8602	0.8011
HBRNN-C	0.8339	0.8088	0.7225

5.6 Hierarchical Attention Bidirectional Recurrent Neural Networks (HA-BiRNN)

In HA-BiRNN [13] comprises of two layers of encoder that are exploited for sentence encoder and word encoder, respectively. Along with this, sentence-level attention and word-level attention are also considered. HA-BiRNN is similar as that of conventional neural network to define. Assume ${}^{A}_{m}h_t^{(n)}$ and ${}^{C}h_t^{(n)}$ as nth layer at time t of mth BiRNN that is exploited for extracting the features and the nth layer at time t of the BiRNN that is exploited for classifying the produced features as shown in Eq. (19). C and A refers to the classification and attribute parameters. As explained above, ${}^{C}W^{kn}$ and ${}^{A}_{m}W^{kn}$ indicate the weight matrix among the nth and k^{th} layers of mth BiRNN and classification RNN.

$$ {}^{A}_{m}h_t^{(n)} = {}^{A}_{m}f^{(n)}\left({}^{A}_{m}W^{(n-1)n}\ {}^{A}_{m}h_t^{(n-1)} + {}^{A}_{m}W^{nn}\ {}^{A}_{m}h_{t-1}^{n} + {}^{A}_{m}b^{(n)}\right) \tag{19} $$

$$ {}^{A}_{m}h_t^{(1)} = {}^{A}_{m}f^{(1)}\left({}^{A}_{m}W^{x1}x + {}^{A}_{m}W^{11}\,{}^{A}_{m}h_{t-1}^{(1)} + {}^{A}_{m}b^{(1)}\right) \tag{20} $$

where ${}^{R}b^{(n)}$ and ${}^{R}f^{(n)}$ indicate the bias vector and the nonlinear activation function of the ${}^{R}h_t^{(n)}$. Equation (20) demonstrates the first layer, input layer as the $(n-1)$th layer and x refers to the input vector. Table 9 demonstrates the comparison analysis of accuracy, recall and precision attained by the HBRNN model over the conventional schemes as per [13].

6 ICD-9 Diagnosis Codes from an Existing EHR Data Repository

"Marshfield Clinic's" have established RDW, which was exploited for analysing breast cancer. The predictive performance of diagnosis entries can be quantified and portrayed in ICD-9 codes. Depending on the hierarchical formation of ICD-9 codes [14] comprising of 3–5 digits, 3 degrees of data depiction is determined: Initial one is level zero that exploits only the initial 3 digits; second one is level 1 that exploits up to the initial 4 digits, and third one is level 2 that exploits up to the entire 5 digits of every code as demonstrated by Fig. 9. The ranking results are given by Table 10 based on [14].

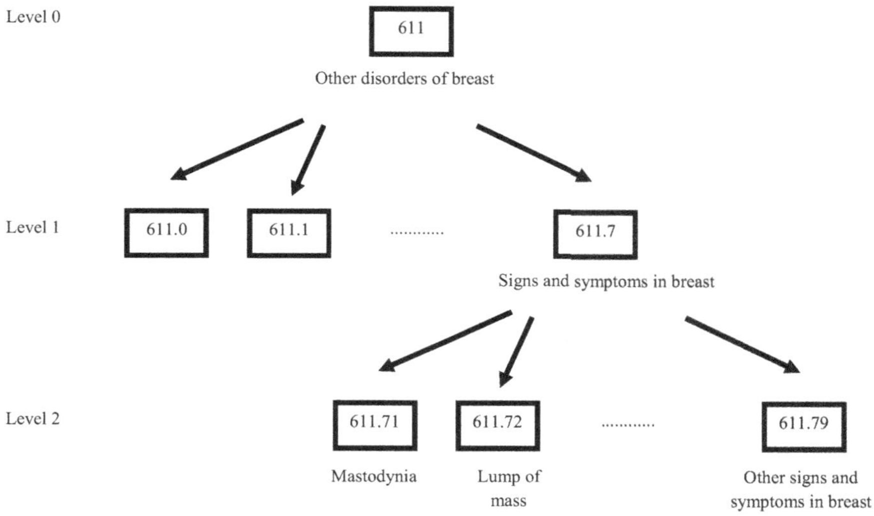

Fig. 9 Levels of EHR data repository

Table 10 Ranking outcomes

LR + Lasso ranking	ICD-9 codes	Description	MI ranking
10	45.13	Other endoscopy of small intestine	21
9	218.1	Intramural leiomyoma of uterus	34
8	367.2	Astigmatism, unspecified	14
7	241.9	Unspecified nontoxic nodular goitre	23
6	V67.9	Unspecified follow-up examination	15
5	610.1	Diffuse cystic mastopathy	6
4	V70.0	Routine general medical examination at a health care facility	3
3	414	Coronary atherosclerosis	4
2	611.72	Lump or mass in breast	2
1	793.8	Abnormal mammogram, unspecified	1

7 Outlier Detection

The outlier detection process indicates a specific classification model: first, the amount of outliers is minute in ratio to the amount of usual instances; and subsequently, the usage of labels in outlier detection is restricted owing to the fact that, the outliers that we are attempting to identify symbolize an unseen or new behaviour. Here, unlabeled data are generally simpler to attain and it indicates the more frequent scenarios in detecting outliers [12]. The most analysed techniques for outlier detection comprise of the following

1. Depth-oriented methods, which are based on computational geometry that computes the different layers of convex hulls and declares the outliers.
2. Distribution-oriented methods that utilize benchmark statistical distribution for modelling the data and it declares the outliers that deviate from the model.
3. Density-oriented techniques, which allocate a weight to every sample depending on their local neighbourhood density.
4. Distance-oriented methods, which evaluate the ratio of database objects that are at a specific distance from a target object (Fig. 10).

In addition, diverse classification is dependent on the outlier detection output and they are categorized into: scoring and labelling techniques. Labelling scheme divide the data into two non-overlying sets (non-outliers and outliers) and the scoring approaches present a ranking list by allocating a factor to every datum reflecting its degree to attain more reliable outcomes [15]. Accordingly, RST is a hypothesis to handle the vagueness, incompleteness and uncertainty and it also overcomes the issues related to classification based on certain similarities [16]. Rough sets were widely exploited for data mining; however, it is rarely adopted for outlier detection in general-domain and for spatiotemporal specific-domain [17].

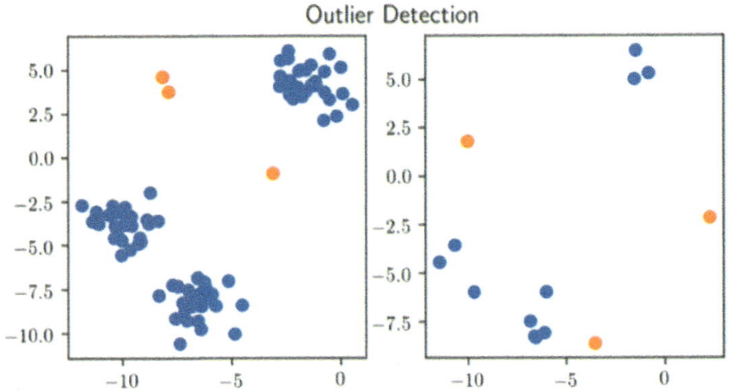

Fig. 10 Outlier detection

8 Conclusions

The various phases for breast cancer detection include pre-processing, feature extraction and classification. Several techniques were deployed for carrying out each stage, by which the accurate decisions can be made regarding the presence or absence of the disease. Moreover, the stage of the tumours such as normal, benign or malignant can also be identified by precise diagnosis. Techniques namely, CLAHE, HM-LCE, temporal filter, mean filter, median filtering, adaptive median filter, Han filter, wiener filter and spatial filters could be exploited for pre-processing the image, whereas, techniques namely SGLDM, gray-map, SFUM, AFUM, PCA, etc. were deployed for extracting the features from the image. Likewise, the breast cancer images could be classified using techniques like SVM, Biclustering and Adaboost Techniques, CNN, RCNN, BI-RADS, HA-BiRNN and so on. In this chapter, a brief review was presented on some of the techniques adopted at each stage (pre-processing, feature extraction and classification) of breast cancer detection. These techniques are helpful to detect the cancer in early stages, so that we can cure the cancer with major probability. Each of the technique can used based on the case of the patient. That is the reason we can use multiple methods for deciding the breast cancer. If we follow these methods we can save many women's lives.

References

1. Sekar B, Lamy J-B, Muro N, Pinedo A, Séroussi B, Larburu N, Guezennec G, Bouaud J, Masero F, Arrue M, Wang H (2018) Intelligent clinical decision support systems for patient-centered healthcare in breast cancer oncology. In: HealthCom 2018, pp 1–6
2. Virmani J, Ravinder A (2019) Effect of despeckle filtering on classification of breast tumors using ultrasound images. Biocybern Biomed Eng 39(2):536–560
3. Berner ES (2009) Clinical decision support systems: state of the art. Agency for Healthcare Research and Quality (AHRQ Publication No. 09-0069-EF). Rockville, Maryland
4. Sahoo AK, Pradhan C, Das H (2020) Performance evaluation of different machine learning methods and deep-learning based convolutional neural network for health decision making. In: Nature inspired computing for data science. Springer, Cham, pp 201–212
5. Das H, Naik B, Behera HS (2020) An experimental analysis of machine learning classification algorithms on biomedical data. In: Proceedings of the 2nd international conference on communication, devices and computing. Springer, Singapore, pp 525–539
6. Bustamam Alhadi, Bachtiar Anas, Sarwinda Devvi (2019) Selecting features subsets based on support vector machine-recursive features elimination and One Dimensional-Naïve Bayes classifier using support vector machines for classification of prostate and breast cancer. Procedia Comput Sci 157:450–458
7. Huang Q, Chen Y, Liu L, Tao D, Li X, On combining biclustering mining and AdaBoost for breast tumor classification. IEEE Trans Knowl Data Eng
8. Ting FF, Tan YJ, Sim KS (2019) Convolutional neural network improvement for breast cancer classification. Expert Syst Appl 120:103–115
9. Alaa AM, Moon KH, Hsu W, van der Schaar M (2016) Confidentcare: a clinical decision support system for personalized breast cancer screening. IEEE Trans Multimed 18(10):1942–1955

10. Yan R, Ren F, Wang Z, Wang L, Zhang F (2019) Breast cancer histopathological image classification using a hybrid deep neural network. Methods (in press, corrected proof, available online 15 June)
11. Koning Jeffrey L, Davenport Katherine P, Poole Patricia S, Kruk Peter G, Grabowski Julia E (2015) Breast imaging-reporting and data system (BI-RADS) classification in 51 excised palpable pediatric breast masses. J Pediatr Surg 50(10):1746–1750
12. Hille H, Vetter M, Hackelöer BJ (2011) The accuracy of BI-RADS classification of breast ultrasound as a first-line imaging method. ISSN 0172-4614
13. Chen D, Qian G, Pan Q (2018) Breast cancer classification with electronic medical records using hierarchical attention bidirectional networks. In: 2018 IEEE international conference on bioinformatics and biomedicine (BIBM). Madrid, Spain, pp 983–988
14. Wu Y et al (2017) Breast cancer risk prediction using electronic health records. In: 2017 IEEE international conference on healthcare informatics (ICHI). Park City, UT, pp 224–228
15. Dey N, Ashour AS, Kalia H, Goswami R, Das H (2019) Histopathological image analysis in medical decision making. IGI Global, Hershey, PA, pp 1–340. https://doi.org/10.4018/978-1-5225-6316-7
16. Khan SU, Islam N, Jan Z, Din IU, Rodrigues JJC (2019) A novel deep learning based framework for the detection and classification of breast cancer using transfer learning. Pattern Recogn Lett 125:1–6
17. Das H, Naik B, Behera HS (2019) Medical disease analysis using neuro-fuzzy with feature extraction model for classification. Inform Med Unlocked 100288

Chapter 4
Energy-Efficient Resource Allocation in Data Centers Using a Hybrid Evolutionary Algorithm

V. Dinesh Reddy, G. R. Gangadharan, G. S. V. R. K. Rao, and Marco Aiello

Abstract Energy efficient scheduling in a cloud data center aims at effectively utilizing available resources and improving the energy utilization. Resource scheduling is NP-hard and often requires substantial computational resources. This chapter presents an interactive PSO-GA algorithm that performs parallel processing of particle swarm optimization (PSO) and genetic algorithm (GA) using multi-threading and shared memory for information exchange to enhance convergence time and global exploration. With the proposed approach, this chapter demonstrates an efficient virtual machine placement in a data center that greatly reduces the total energy consumption up to 34% and the convergence time up to 50% compared to PSO, GA, and modified best-fit decreasing approaches. Further, this method achieves average parallel efficiency of 90% and a speed up of 1.5. This chapter also evaluates effectiveness of the proposed algorithm on benchmark optimization test problems over the state-of-the art algorithms.

1 Introduction

Cloud is the collective name for a broad collection of services for transparent on-demand network access to data center resources. Organizations can choose where, when, and how they use these services based on the pay-as-you-go model [1]. Cloud

V. D. Reddy
Cognizant Technology Solutions, Hyderabad, India
e-mail: dineshvemula@gmail.com

G. R. Gangadharan (✉)
National Institute of Technology, Tiruchirappalli, India
e-mail: geeyaar@gmail.com

G. S. V. R. K. Rao
Cognizant Technology Solutions, Chennai, India
e-mail: gsvrkrao@gmail.com

M. Aiello
Institute of Architecture of Application Systems, University of Stuttgart, Stuttgart, Germany
e-mail: aiellom@ieee.org

J. K. Rout et al. (eds.), *Machine Learning for Intelligent Decision Science*,
Algorithms for Intelligent Systems, https://doi.org/10.1007/978-981-15-3689-2_4

data centers enable its users to utilize computing resources as a service, rather than owning them [2, 3]. To serve the user's request, the server needs to provide all the resources required by a virtual machine (VM) including hard disk, RAM, bandwidth, CPU, etc.

Cloud-based solutions are constantly gaining popularity and more data center infrastructure is being required to support their offerings. Increasing cooling demand and power cost can be a huge drain on operating budgets of high-density data center environments. Researchers have been working to improve the energy efficiency of data centers through various practices. From the IT perspective, efficient VM placement techniques, decommissioning of unused servers, and consolidation are considered the best ways to save energy in data centers. VM placement is an important task in which one selects a proper physical machine for hosting the virtual machine. In light of its importance, researchers proposed various algorithms for VM placement in data centers. Each of these algorithms applies some placement factor to achieve the objectives of the problem.

Data center consumers would like to provision resources transparently with minimal latency. From the perspective of users, the performance of the data center is measured in terms of response time, VM provisioning time, service provision time, etc. To serve the customers in a better way, data center providers should follow an optimal VM provisioning within very short time. Traditional algorithms like Best-Fit, First-Fit, Modified Best-Fit, etc. will not be efficient for large scale data centers as they do not provide the optimal placement. PSO finds the near-optimal solution in an acceptable time whereas GA may not provide the optimal solution as fast as PSO. These algorithms performed well in terms of achieving near-optimal solutions, but we observed that they require a lot of computational resources, taking much time to converge and require improvement in optimal solution. PSO can't work out for the problem of scattering and can converge prematurely. ACO, simulated annealing (SA), and firefly algorithms [4–6] are also used for virtual machine placement. But in ACO, probability distribution can change for each iteration and convergence time is uncertain. In simulated annealing, repeated annealing with a $1/logk$ schedule may take long time for expensive objective functions. It takes a lot of time to find optimal solutions. Firefly algorithm parameters will not change over time which may degrade the exploration [7]. Firefly algorithm is overly restrictive for high dimensional and nonlinear problem. This method does not have the mechanism to memorize the best solution found [8].

For solving these problems, this work proposes an "Interactive PSO-GA" for VM placement, combining the searching abilities of both GA and PSO. To make particles more suitable to the environment before producing offsprings, we incorporate the social interaction between the algorithms to enhance the population on each generation. In our algorithm, basic PSO and GA algorithms run in parallel in separate threads. After each iteration, these threads share the top ranked particles information with each other to form the population for the next iteration and override the poor-performing particles. The proposed algorithm inherits a natural parallelism and avoids the problem of performance deterioration of PSO and GA with increasing dimension. Our approach dynamically chooses the particles between GA and PSO to improve the solution quality.

The remainder of the chapter is organized as follows. Section 2 presents the review of several virtual machine placement algorithms. The proposed IPSOGA algorithm, details of calculating speed up and parallel efficiency are given in Sect. 3. Section 4 presents the experimental evaluation of energy-aware VM allocation and benchmark functions evaluation using IPSOGA followed by concluding remarks in Sect. 5.

2 Related Work

As cloud resource provisioning in data centers is considered as a NP hard problem [9, 10], soft computing techniques are applied by researchers to solve the problem of virtual machine placement and selection, to find the optimal solution with reduced cost and computational time, satisfying the service level agreement. Portaluri et al. [11] proposed energy-aware dynamic allocations of virtual machines considering computational and network requirements using multi-resource Best-fit algorithm.

Lee et al. [12] described a VM consolidation method considering the competing resource demands in multiple dimensions. They developed two heuristic approaches, namely, Dot-Product and Norm-based Greedy to select the next VM. Wang et al. [13] presented a heuristic min-cost algorithm to solve energy consumption problems in data centers, resolving difficulties with integer decision variables and non-linearity of the power model. This approach takes more time in large scale scenarios. Sayad-navard et al. [14] proposed a Markov chain model to predict the reliability of each server and developed algorithms for consolidation process to reduce the number of active servers considering the reliability and CPU utilization. They are able to avoid inefficient VM migrations and reduce energy consumption. Shaw et al. [15] developed an anti-correlated placement algorithm considering CPU and bandwidth consumption to predict the resource demand.

As the VM placement issue is an NP-hard problem [16, 17], the deterministic algorithms may not always give the global optima. Bio-inspired approaches can provide global optimal solutions [18, 19]. These approaches have used for classification and feature extraction in medical disease analysis [20–23]. Further, these approaches are used in health recommendation system [24], exchange rates forecasting [25]. Genetic algorithms encode the solutions in a chromosome-like data structure and perform the evolutionary operations for finding global optimal solution. Since they evaluate many points simultaneously, they are efficient at exploring the search space completely. PSO is a swarm-based evolutionary computing technique that accomplishes the global optimization in a faster way. PSO mimics the social behavior of the insects and uses swarm theory. Particles gain velocity on their own and from neighborhood flying experiences. Compared to PSO, GA is relatively poor in finding the global optimum solution for complex problems. PSO has constructive co-operation between particles that improve the learning performance and convergence rate of the algorithm.

Mohanty et al. [26] proposed novel group-based job scheduling algorithm to provide users with high response time in cloud environment. This algorithm groups independent jobs considering the communication time. Gharehpasha et al. [27] proposed a hybrid discrete multi-object sine cosine algorithm and multi-verse optimizer for optimal placement. Their goal is to reduce the power consumption and resource wastage. Grange et al. [28] proposed an attractiveness-based blind scheduling algorithm using a greedy heuristic approach for choosing a definitive placement for a task at the time of its submission. This algorithm compares multiple possible placements. Gupta et al. [29] proposed a resource usage factor for VM placement to improve the resource utilization. Further, they developed a new resource usage model for detecting unbalanced utilization of resources on the active physical machines. Zhiyong et al. [30] proposed a chemical reaction optimization to solve VM placement problem. Zhao et al. [4] explored the balance between server power savings and virtual machine performance, using ant colony optimization (ACO). Liu et al. [5] proposed ACO algorithm for virtual machine placement, combining ACO with a local search technique. This method effectively minimizes the energy consumption by reducing the active physical machines.

The combination of GA and PSO algorithms has been investigated in many studies [31–34]. However, to the best of our knowledge, parallelization of these algorithms is not considered in the literature instead authors worked on serialization of these algorithms in one or other ways [35, 36]. Parallelization helps to enhance the computational throughput and global search capability. If the algorithm is properly designed to take full advantage of the multi-core architecture, it can execute faster than their serial counterparts. So, by taking the recent advances in multi-core architectures and using multi-threading and shared memory, developed a parallelized optimization algorithm called "Interactive PSO-GA" (IPSOGA). This algorithm performs parallel processing of PSO and GA using multi-threading. This helps in balancing between improving convergence time and accuracy.

3 Interactive PSO-GA

3.1 Particle Swarm Optimization (PSO)

PSO is a swarm-based evolutionary computing technique that accomplishes the global optimization in a faster way [37, 38]. Particles gain velocity on their own and from neighborhood flying experiences. In addition, PSO can memorize the best solutions after each iteration (see Fig. 1).

PSO has stochastic mechanism that makes the particles to exploit some specific areas and rise up from the local optima. PSO has ability to keep track the particle that has best value in the population (L^j) and this is global best. Best experience each particle had till that point is called particle's local best (L_i). Equations 1 and 2 represent the updated particle velocity and position, respectively.

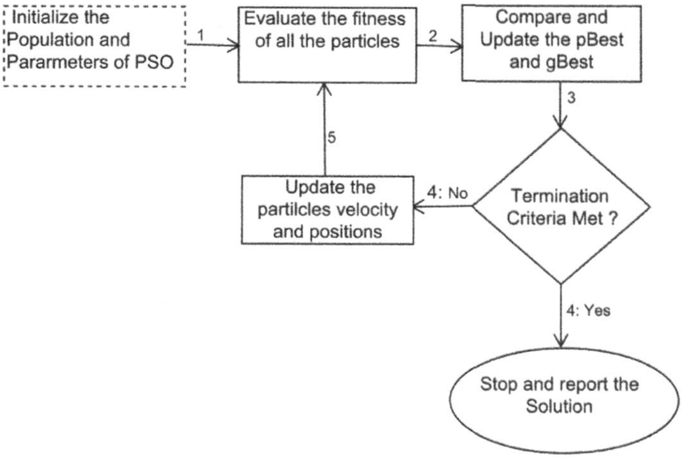

Fig. 1 Flow of particle swarm optimization

$$\text{Velocity}_i(t+1) = w\text{Velocity}_i(t) + k_1 r_1(\text{Local best} - X_i(t)) + k_2 r_2(\text{Global best} - X_i(t)) \quad (1)$$

$$X_i(t+1) = \text{Velocity}_i(t+1) + X_i(t) \quad (2)$$

where k_1 and k_2 are the learning factors, w is the inertia weight, and r_1 and r_2 are the random numbers between (0,1).

3.2 Genetic Algorithm (GA)

A genetic algorithm is a metaheuristic algorithm based on the process of natural selection [39]. Each individual of the population is coded to make a chromosome that represents a candidate solution for the given optimization problem.

Individuals who are more successful in adapting to their environment are selected based on their cumulative probability calculated based on fitness. The selected individuals reproduce the offsprings by exchanging pieces of their genetic information (characteristics of parents) depending on the cross over rate, whilst individuals who are less fit will be eliminated. This phase is known as crossover. Mutation rate is a measure of the likeness that random elements of your chromosome will form a specific mutant.

To improve the "offspring" solutions, mutation operator is applied by altering some genes in the strings depending on the mutation rate. Swap mutation is used in our approach where it randomly chooses two genes on the chromosome and swaps those values. This selection-crossover-mutation cycle repeats until a satisfactory solution is found as shown in Fig. 2.

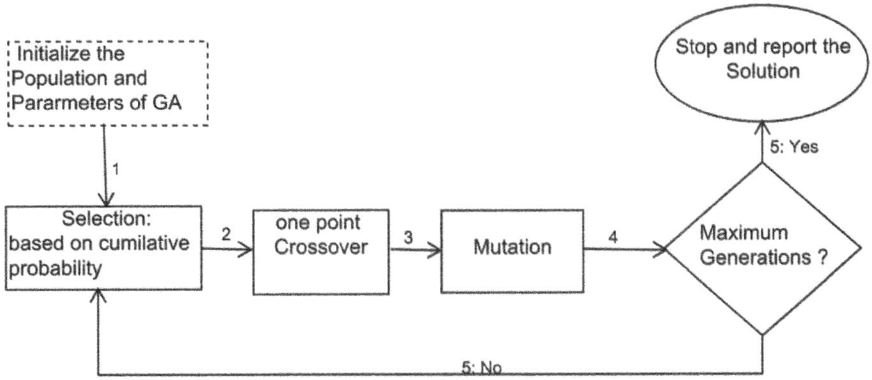

Fig. 2 Flow of genetic algorithm

3.3 Modeling Energy-Aware VM Allocation

Let $(v_1, v_2, v_3, \ldots, v_m)$ be the set of VMs to be placed in the servers $(p_1, p_2, \ldots, p_n)$ of the data centers. Then, we define each particle position as a list of servers as shown in Eq. 3. $X_i(t)$ gives corresponding placement for each virtual machine for a time period (t).

$$X_i(t) = (p_1, p_5, p_1, \ldots, p_j, \ldots, p_n, \ldots, p_j) \tag{3}$$

The allocation must satisfy the following constraints:

- Each VM has to be placed in only one physical server.
- Resource requests like memory, bandwidth, and cpu cycles of virtual machines allocated in a server should not exceed the capacity of the server.

Our objective is to maximize the tuple as follows:

$$\text{Maximize} \left\{ \frac{v_i^{\text{cpu}}}{p_j^{\text{cpu}}}, \frac{v_i^{\text{mem}}}{p_j^{\text{mem}}}, \frac{v_i^{\text{bw}}}{p_j^{\text{bw}}} \right\}$$

where $v_i^{\text{cpu}}/p_j^{\text{cpu}}$ gives the cpu utilization of ith virtual machine (v_i) in the jth host (p_j). $v_i^{\text{mem}}/p_j^{\text{mem}}$ gives the memory utilization of ith virtual machine (v_i) in the jth host (p_j), and $v_i^{\text{bw}}/p_j^{\text{bw}}$ gives the bandwidth utilization of ith virtual machine (v_i) in the jth host (p_j).

This chapter aims to minimize the energy consumption of the host $(E(p_i))$ when placing virtual machines while satisfying service level agreements. The power consumption of a server (p_i) is calculated based on the CPU utilization (u_i) according to Eq. 4.

$$E(P_i) = \text{Idle}_i \ . \ e_i^{\max} + (1 - \text{Idle}_i) \ . \ e_i^{\max} \ . \ u_i \tag{4}$$

where Idle_i and $e_i^{\max}$ are the power consumption of a server p_i when it is active idle and fully utilized, respectively, [40].

The aim is to minimize energy consumption of the data center defined in Eq. 5.

$$\text{Fitness function}\,(f) = \sum_{k=1}^{n} E(P_k) \tag{5}$$

3.4 Interactive PSO-GA (IPSOGA)

Our algorithm starts with randomly initializing the population and other parameters like inertia coefficients, learning factors, crossover rate, and mutation rate. This algorithm calculates the fitness value of every solution using Eq. 5 and choose the global best that has the maximum fitness value.

With the said information, the program is divided into two discrete parts that can run concurrently in two different threads (see Fig. 3). The first thread runs the PSO algorithm and the second thread runs GA algorithm. The first thread follows the flow explained in Sect. 3.1 where individual coordinates of each particle get closer to the global best of the population. The particle updates its velocity using local best, global

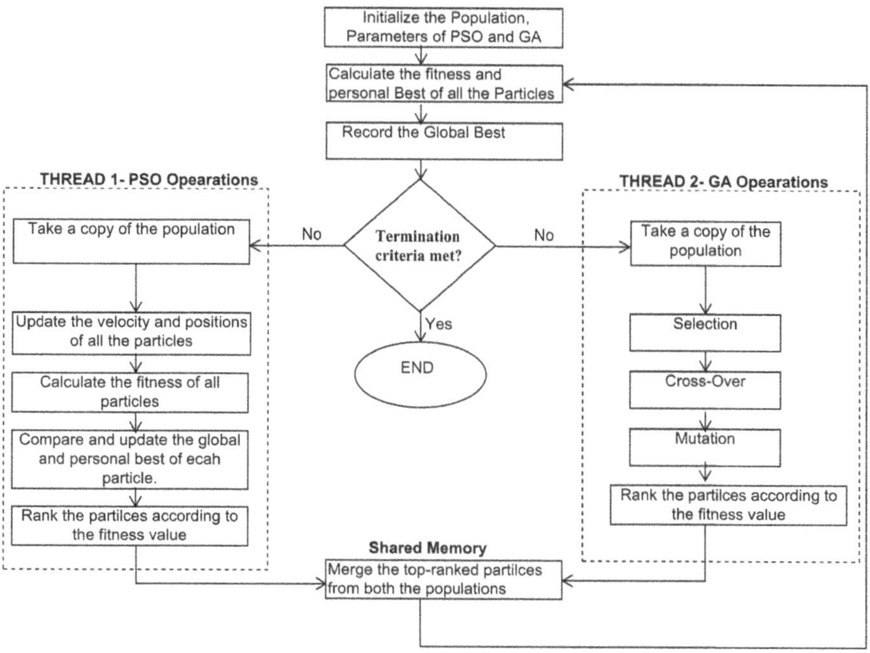

Fig. 3 Flow diagram of IPSOGA

best, and initial velocity as given in Eq. 1. Further, the position of each particle is updated using the updated velocity, and its current position as shown in Eq. 2.

The second thread follows the flow as given in Sect. 3.2. A chromosome defines a proposed solution to the VM placement problem which represents a list of servers (randomly selected). The length of each chromosome is equal to the number of VM requests at that instance. In this chapter, roulette wheel selection is used for selecting potentially useful solutions for reproduction. The fitness score is associated with individual chromosome that indicates how close a chromosome is to the optimal solution. The selection probability (Prob(i)) of a chromosome in a population is defined in Eq. 6. We have used single-point crossover for merging the genetic information of two individuals.

$$\text{Prob}(i) = \frac{f_i}{\sum_{i=1}^{n} f_i} \tag{6}$$

After each iteration, both the threads rank the particles of the respective population according to the particles fitness value. We create a new population for the next iteration by merging the top particles (according to fitness) in shared memory context as shown in Algorithm 2, where the process of merging is done asynchronously. Individual particles in PSO and GA are updated according to the operations given in Sects. 3.1 and 3.2. This process continues till the maximum number of iterations reached or till we observe s-continuous improvements as shown in Algorithm 1.

```
Initialize each particle position randomly within the permissible range.
foreach particle X_i do
    Calculate the fitness of the partcle.
    Store the local best of the particle.
end
L̂_j(t)= particle with maximum fitness value in the swarm. // Global Best
repeat
    Perform the PSO operations and GA operations in different threads.
    Rank the indiviual populations of PSO and GA according to the fitness values.
    newPopulation = mergePopulation(psoPopulation, gaPopulation);
    L̂_j(t + 1)= particle with maximum fitness in the population.
    if fitness (L̂_j(t + 1)) > fitness(L̂_j(t)) then
        L̂_j(t) = L̂_j(t + 1).
        Boolean Improved = True;
        ImpCount ++;
        Update the populations of the PSO and GA with the newPopulation.
        if Improved && ImpCount > s then
            Return L̂_j(t).
        else
            Improved = false.
            ImpCount=0;
        end
until maximum-iterations;
```

Algorithm 1: IPSOGA

```
input  : population1, population2
output: newPopulation
temp1= first particle of the population1;
temp2= first particle of the population2;
p=g=0;
for i=0; i<population1.length(); i++ do
    if fitness(temp1) ≥ fitness(temp2) then
        newPopulation[i]=population1[p];
        temp1= population1[++p];
    else
        newPopulation[i]=population2[g];
        temp2= population2[++g];
    end
end
return newPopulation
```

Algorithm 2: mergePopulation()

On a multi-core machine, we can improve the actual performance of compute-bound operations, by executing multiple threads in parallel, with every core handling a separate thread simultaneously. Evolutionary algorithms such as PSO and GA are better suited for parallel implementation. In both the algorithms, population size can be adjusted according to the availability and capacity of processors. However, increasing the population size allows for a higher global exploration in the search space and avoids the chance of trapping into local minima.

The basic PSO algorithm was designed in synchronous fashion for updating particle best (pBest) and swarm's best (gBest) fitness values along with their positions. In this approach, we evaluate the fitness of all the particles before updating the pBest and gBest values. In our approach, we evaluate the fitness values concurrently, which indicates the synchronous design. However, these values can be updated in an asynchronous manner [41, 42]. The parallel implementation requires updating gBest of the population after PSO and GA complete their updates. Consequently, this will reduce the convergence rate compared to the asynchronous implementation. To improve the exploration capability, we perform the enhancement of the population using the ranked particles from both PSO and GA algorithms. After each iteration, we rank the individuals of the populations concurrently. We update the initial population with the top-ranked individuals chosen from both the populations, as shown in Algorithm 2. There is no direct decency between GA and PSO except the population update.

The fundamental reason for parallelizing PSO and GA programs is to make them interactable and run quicker. To evaluate the advantage of parallel program, we have used the speedup parameter. The speedup of a parallel code refers to how much faster it runs in parallel. Speedup expresses the relative speed of the parallel program compared to the sequential one [43].

$$\text{Speedup} = \frac{\text{Time for sequential run } (T_s)}{\text{Time for parallel run } (T_p)}$$

where T_s is used to express the execution time for the given optimization problem when PSO and GA run in sequence. We took the fastest execution time among 20 runs. T_p is the execution time of the parallelized version of the program (IPSOGA) for solving the same problem.

If multiple threads are executed individually on a computer with p processors, the speedup is equal to p. But the maximum speedup is achievable, only when the load is balanced among the available processors and with no communication cost. In such an idle environment, each processor requires T_s/p time units to complete the thread execution and the speedup is given as follows:

$$\text{Speedup} = \frac{T_s}{T_s/p} = p$$

However, according to the Amdahl law [44], it is difficult to obtain a speedup value "p" because every program has a fraction (δ) of code that cannot be parallelized. In our case the code for merging two populations after each iteration cannot be parallelized. The remaining code can be executed in parallel. The execution time and the speedup is given as follows:

$$T_p = T_s * \delta + T_s * (1 - \delta)/p$$

$$\text{Speedup} = \frac{T_s}{T_p} = \frac{p}{\delta * (p - 1) + 1}$$

4 Experiments and Performance Evaluation

We resort to the CloudSim framework [45] to model and simulate the real configurations of cloud computing infrastructure. The experimental setup simulates a data center that comprises 100 servers, which are characterized as shown in Table 1. These servers include G4 Xeon 3040, G4 Xeon 3075, G3 Pentium D 930 and are homogeneous. Table 2 presents the virtual machines configurations. Virtual Machine Provisioner is responsible for allocating VMs to physical nodes. Resource requests are of heterogeneous nature and VM parameters include Memory, CPU, Network, and Disk. The number of hosts is constant in every experiment where as VMs differ

Table 1 Physical machines configurations

Host name	CPU	Memory (GB)	Network (GBPS)	Disk (GB)
G4 Xeon 3040	3000	8.5	30	125
G4 Xeon 3075	3000	8.5	30	125
G3 Pentium D 930	3000	8.5	30	125

Table 2 Virtual machines requirements

VM type	CPU	Memory (MB)	Network (MBPS)	Disk (GB)
Type 1	100	613	100	2.5
Type 2	200	870	100	2.5
Type 3	300	1740	100	2.5

in between from 50 to 300. The requirements for the Vms are produced arbitrarily. Our proposed IPSOGA is integrated into CloudSim using a personal computer with Intel Core i7-5500U, 2.40 GHz Dell Latitude with four cores, and 8 GB of RAM.

We performed several experiments on the Cloudsim to initialize the parameters for IPSOGA and other algorithms. The probabilities of crossover and mutation are initialized to 0.6 and 0.05, respectively. In PSO [46], the parameters were set as follows: learning factors (k1, k2) are set to (2, 2). Inertia weight is initialized to 1 and it is decreased iteratively by 0.05 to cause the particles to slow down and converge a bit easier. After several experiments, we initialize the population size to 40 for linear convergence rate and maximum-iterations to 150 for ensuring global exploration.

For IPSOGA approach, the number of threads used for each evaluation is set to 2 and all the approaches are allowed to run until there are '3' continuous improvements in the optimal solution or 150 function evaluations are completed.

4.1 Performance Analysis in Terms of Energy Consumption

We compare our proposed IPSOGA algorithm with Particle Swarm Optimization [19], Genetic Algorithm (GA) [18], and Modified Best-Fit Decreasing (MBFD) allocation [40]. MBFD first sorts all the virtual machines in decreasing order of their utilization and allocates a VM to the host where energy consumption will be minimum [40]. For 300 VMs, our IPSOGA approach consumes the minimum energy of 2.27 kWh where as PSO, GA, and MBFD consume 2.36 kWh, 2.75 kWh, and 3.25 kWh, respectively. PSO and GA are able to decrease the energy consumption by reducing the number of active servers. However, they fall mostly into local maxima due to lack of exploration capability after certain extent. We observe that our proposed approach is able to reduce the number of active servers thereby reducing the energy consumption by generating new population using the best candidates from both the populations. On average, IPSOGA decreases the energy consumption up to 17% over GA, 10% over PSO and 34% over MBFD policies. We repeated the experiments on the same cloud setup by increasing the user requests (VMs) from 50 and going to 300 and the energy consumption of the data center for various approaches are given in Table 3 (Fig. 4).

Table 3 Analysis of energy consumption for various placement algorithms

	Energy consumption (kWh)			
VMs	MBFD	PSO	GA	IPSOGA
50	1.53	1.01	1.21	0.91
100	1.75	1.47	1.49	1.01
150	2.08	1.64	1.72	1.41
200	2.33	1.85	1.75	1.66
250	2.92	1.88	2.12	1.87
300	3.25	2.36	2.75	2.27

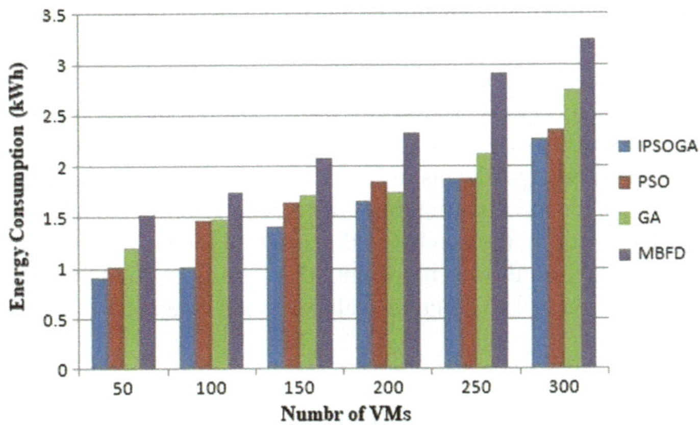

Fig. 4 Energy consumption comparison

4.2 *Performance Analysis in Terms of Convergence*

For metaheuristic approaches, the complexity is specified in terms of the hardness of the problem and the size of input than the algorithm because the complexity of the algorithm is affected by the representation, initialization of parameters, termination condition and the diversity of the initial solutions, etc. For each algorithm, the convergence rate is calculated as the average number of iterations it takes to reach the global optima. The convergence graph of the IPSOGA algorithm for varying input requests (VMs), ranging from 50 to 300 given in Fig. 5. We compared the proposed algorithm with PSOGA algorithm. In PSOGA model the initial population of PSO is created with output of GA [35, 47].

From the results, we show that our approach performs well when increasing the number of virtual machines. IPSOGA decreases the convergence time up to 50% compared to PSOGA. Further, it shows lower convergence rate with the increasing of design variables. The convergence time of IPSOGA and PSOGA for varying virtual machines is given in Fig. 5.

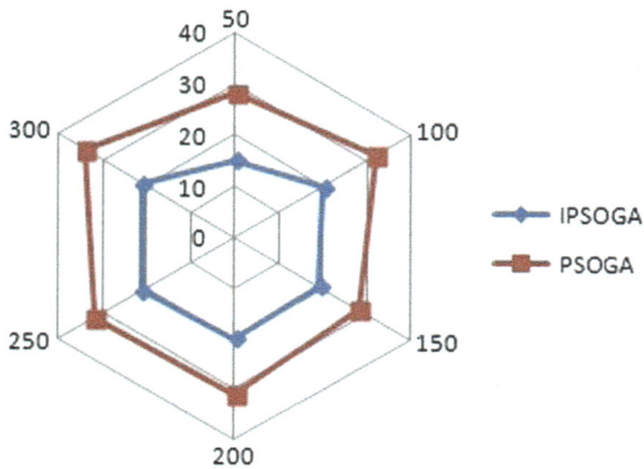

Fig. 5 Convergence analysis

4.3 Performance Analysis in Terms of Speedup and Parallel Efficiency

Parallel performance is calculated using the speedup as given in Sect. 3. We use the sequential execution time of PSO, GA and the execution time of the IPSOGA for calculating speedup. In general, parallel efficiency is the ratio of speedup to the number of threads used. For the virtual machine placement problem, we reported the speedup for 50, 100, 150, 200, 250, and 300 design variables.

We solved the virtual machine placement problem using 50 particles, 2 threads and allowed to run until there are *r* continuous improvements in optimal solution or 150 fitness evaluations are completed. The execution time of IPSOGA for solving the VM placement problem along with Parallel efficiency is described in Table 4.

Table 4 Speedup and parallel efficiency of IPSOGA

	IPSOGA	
Number of VMs	Parallel efficiency (%)	Speedup
50	97.5	1.95
100	100	2
150	87.5	1.75
200	82.5	1.65
250	85	1.70
300	80	1.60

Table 5 Benchmark functions and their description

Function	Dim	Range	fmin
$f_1(\mathrm{x}) = f(x_1, \ldots, x_n) = \max_i \{\lvert x_i \rvert, 1 \le i \le n\}$	30	[−100, 100]	0
Six-humb camel back Function $f_2(\mathrm{x}) = f(x_1, x_2) = (4 - 2.1x_1^2 + \frac{x_1^4}{3})x_1^2 + x_1x_2 + (-4 + 4x_2^2)x_2^2;$	2	−3 <= x1 <= 3, −2 <= x2<= 2	−1.0316
Quartic Function $f_3(\mathrm{x}) = f(x_1, \ldots, x_n) = \sum_{i=1}^{n} i x_i^4 + \mathrm{random}(0, 1)$	30	[−1.28, 1.28]	0
Schwefel Function $f_4(\mathrm{x}) = f(x_1, x_2, \ldots, x_n) = 418.9829d - \sum_{i=1}^{n} x_i \sin\left(\sqrt{\lvert x_i \rvert}\right)$	30	[−100, 100]	0
Ackley Function $f_5(\mathrm{x}) = f(x_1, \ldots, x_n) = -20.\exp\left(-0.2\sqrt{\frac{1}{n}\sum_{i=1}^{n} x_i^2}\right) - \exp\left(\frac{1}{n}\sum_{i=1}^{n}\cos(cx_i)\right) + 20 + \exp(1)$	30	[−32, 32]	0
Fifth Dejong Function $f_6(\mathrm{x}) = f(x_1, \ldots, x_n) = \left(0.002 + \sum_{i=1}^{25} \frac{1}{i + (x_1 - a_{1i})^6 + (x_2 - a_{2i})^6}\right)^{-1}$ where $a = \begin{bmatrix} -32 & -16 & 0 & 16 & 32 & -32 & \ldots\ldots\ldots & 0 & 16 & 32 \\ -32 & -32 & -32 & -32 & -32 & -16 & \ldots\ldots\ldots & 32 & 32 & 32 \end{bmatrix}$	2	[−65.536, 65.536]	0
Dejong's Function 1 $f_7(\mathrm{x}) = f(x_1, \ldots, x_n) = \sum_{i=1}^{n} x_i^2$	30	[−5.12, 5.12]	0
Beale Function $f_8(\mathrm{x}) = f(x_1, x_2) = 100\left(x_2 - x_1^3\right)^2 + (1 - x_1)^2$	2	[−4.5, 4.5]	0
Leon Function $f_9(\mathrm{x}) = f(x_1, x_2) = (1.5 - x_1 + x_1x_2)^2 + \left(2.25 - x_1 + x_1x_2^2\right)^2 + \left(2.625 - x_1 + x_1x_2^3\right)^2$	2	[0, 10]	0
Function $f_{10}(\mathrm{x}) = f(x_1, \ldots, x_n) = \frac{1}{2}\sum_{i=1}^{n}\left(x_i^4 - 16x_i^2 + 5x_i\right)$	30	[−5, 5]	0

4.4 Validation Against Benchmark Test Problems

We have evaluated our IPSOGA approach against some of the well-known benchmark test functions including multi-modal and uni-modal [48–50] functions. The functions listed in Table 5 are some of the common functions used for testing optimization algorithms.

These benchmark results have been compared with the standard meta-heuristics such as PSO [46], GA [18], Modified Discrete PSO (MDPSO) [51], and PSO using eagle strategy (ESPSO) [52]. In our experiments, we have initialized the population size for each algorithm as 40. We used a population of 50 particles, and these methods are allowed to run until the target is reached or 10000 function evaluations are completed. For comparison, we consider the average fitness value and standard deviation for 30 epochs. These statistical results are presented in Tables 6 and 7 respectively.

Table 6 Comparative performance of average fitness value for benchmark functions

	PSO	IPSOGA	GA	ESPSO	DPSO
F1	44.5689	39.3632	80.7633	80.1867	51.777
F2	−1.03128	−1.03165	−1.03154	−1.03101	−1.0313
F3	9.56122	7.11337	18.53261	4.79054	12.8333
F4	7418.58	7646.08	9071.25	8742.45	9336.38
F5	0.2064	0.09517	0.23875	0.203	0.29468
F6	1.01942	0.99858	0.99831	0.99844	0.99841
F7	44.56894	39.36316	80.76326	80.18667	58.1227
F8	0.10404	0.00011	1.18203	0	0.10078
F9	0.00228	0.00802	0.03865	0.02953	0.02007
F10	0.00255	0.00225	0.52118	0.00404	0.00552

Table 7 Comparative performance of standard deviation for benchmark functions

	PSO	IPSOGA	GA	ESPSO	DPSO
F1	3.0961	2.6203	3.2677	2.5403	3.793
F2	0.00022	0.00014	0.00005	0.00108	0.00019
F3	2.9398	0.9541	2.7818	0.3624	3.8268
F4	430.0222	199.9183	227.2061	291.9336	190.8128
F5	0.1837	0.07379	0.13970	0.18129	0.21080
F6	0.05227	0.00059	0.00049	0.00056	0.00003
F7	14.7862	4.3835	5.7622	5.2452	13.1623
F8	0.25748	0.00012	1.5072	0	0.18044
F9	0.00685	0.00771	0.12523	0.02597	0.0507
F10	0.00293	0.00204	0.39337	0.00323	0.00448

For MDPSO, we initialize k1 with 3, and k2 with 2, respectively [51]. For ESPSO, c1, c2 are set to 2 and wmin, wmax, λ are set to 0.4, 0.9, and 2, respectively [52]. For Genetic Algorithm and PSO the parameters are initialized as described in the beginning of Sect. 4.

The convergence curves for benchmark functions (F1–F10) using various algorithms are shown in Figs. 6, 7, 8, 9, 10, 11, 12, 13, 14 and 15. Among the chosen benchmark functions, F1, F4, F7 are uni-modal functions and F2, F3, F5, F6, F8, F9, F10 are multi-modal functions. From the results, we can see that the IPSOGA method is able to escape local optima and converges to near optimum solution faster than the other algorithms.

Figures 6, 9, and 12 inquire into exploitation capabilities of the various algorithms. IPSOGA has shown good performance compared to traditional algorithms. IPSOGA is able to reach the global optimum faster for F1 and F7. From Figs. 7,13,14 and 15, it is clear that GA, ESPSO are stuck in local optima for most of the multi-modal

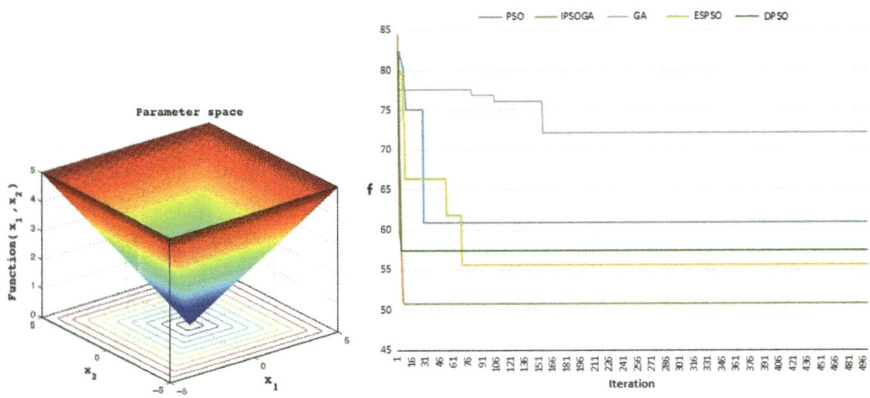

Fig. 6 Contour and convergence for (F1)

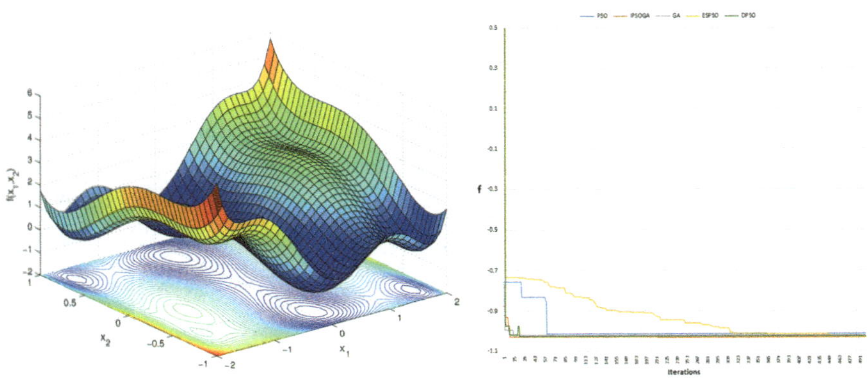

Fig. 7 Contour and convergence for six-hump camel back function (F2)

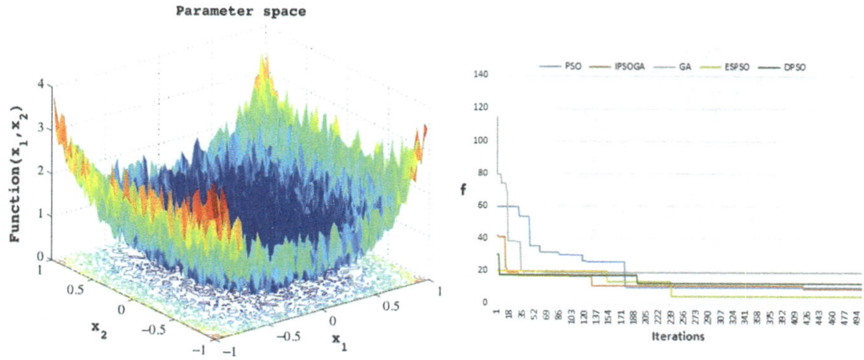

Fig. 8 Contour and convergence for quartic function (F3)

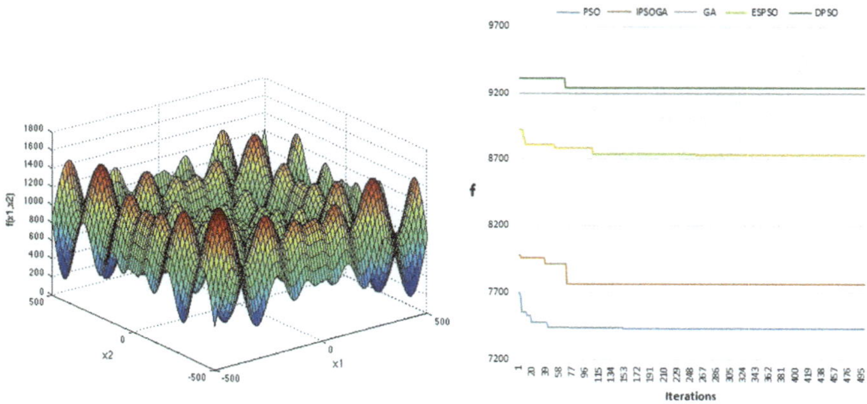

Fig. 9 Contour and convergence for schwefel function (F4)

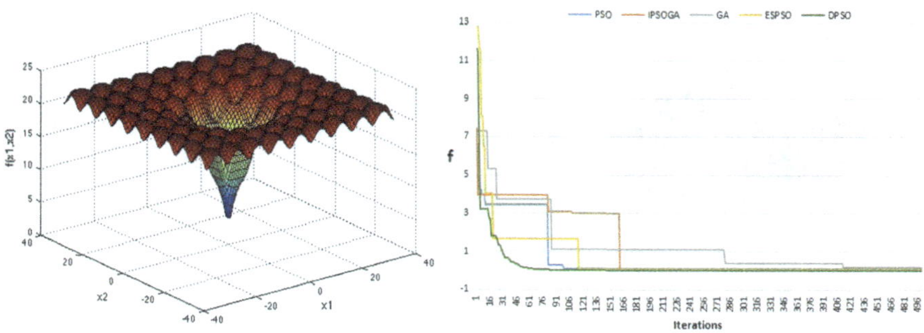

Fig. 10 Contour and convergence for ackley function (F5)

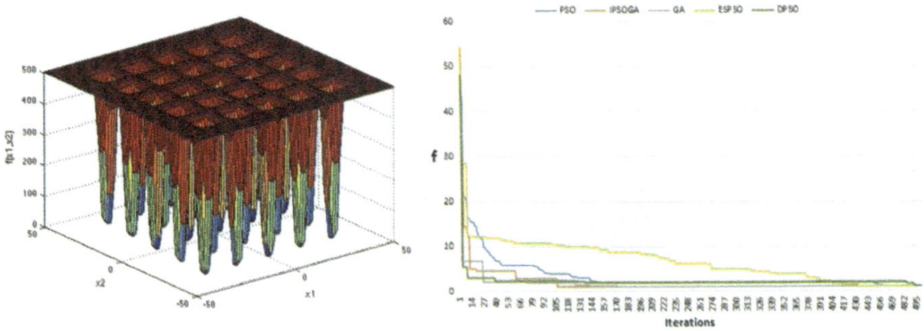

Fig. 11 Contour and convergence for fifth dejong function (F6)

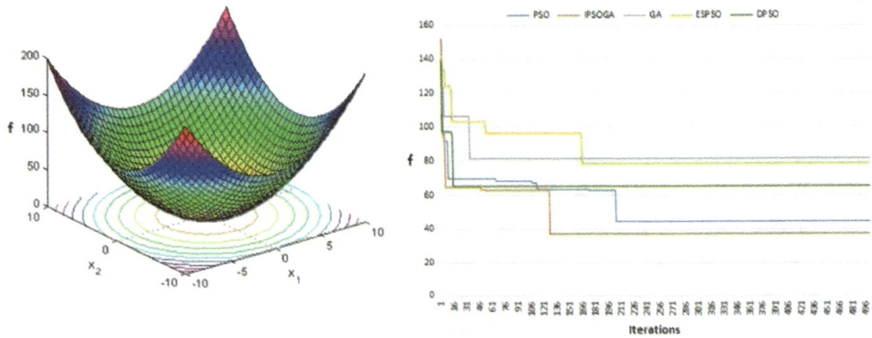

Fig. 12 Contour and convergence for Dejong function1 (F7)

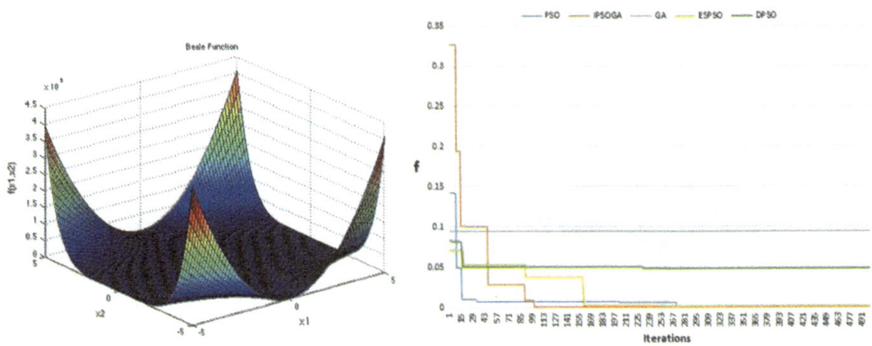

Fig. 13 Contour and convergence for beale function (F8)

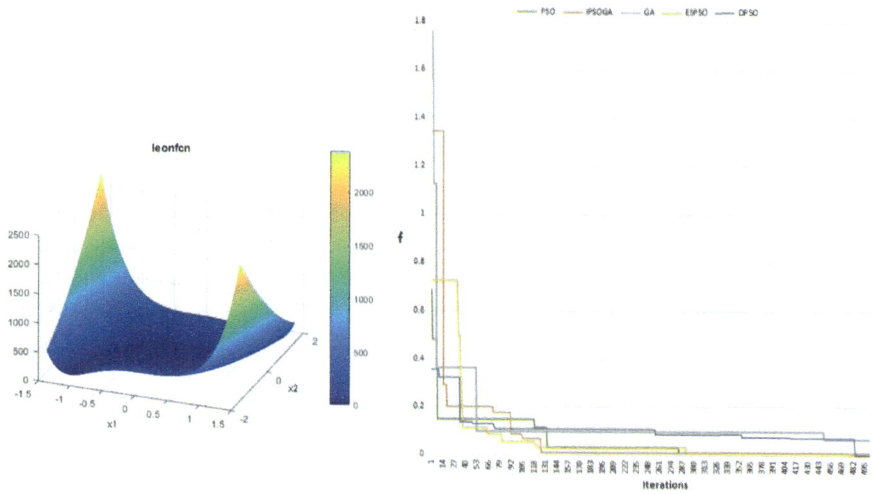

Fig. 14 Contour and convergence for Leon function (F9)

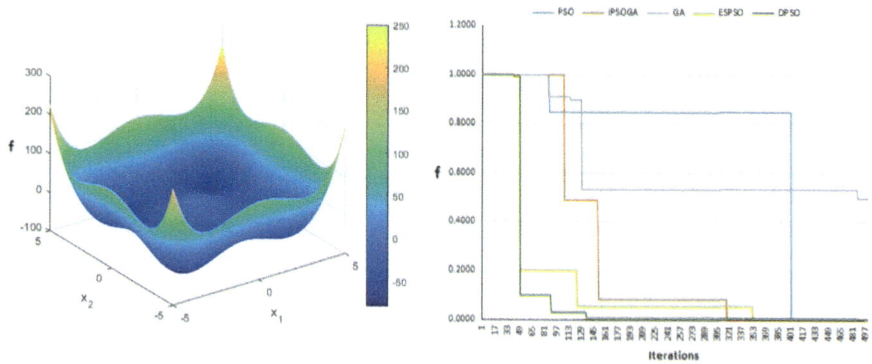

Fig. 15 Contour and convergence for F10

functions where as IPSOGA is able to explore more and reach global optima for multi-modal functions. As IPSOGA considers the high quality particles in each iteration, it improves the diversity of population and helps in reaching global optimum faster.

5 Conclusion

The parallel implementation of "Interactive PSO-GA" algorithm presented in this study exhibits excellent parallel performance and reduces the convergence time. We used multi-threading and shared memory to run the algorithms in parallel that communicate with each other. We used this algorithm to solve energy-efficient resource allocation in cloud data centers. A systematic investigation of parallel performance and scalability for a IPSOGA algorithm has been presented. Our method has demonstrated a significant improvement in convergence time up to 50%, reduced energy consumption up to 34% and achieved a speed up of 1.5. Further, we evaluated the algorithm on benchmark test functions and our results indicate the high stability and scalability.

Though IPSOGA achieves parallel efficiency, it has some common disadvantages. It is complex to debug and test the processes as it is much harder to replicate an error in a multi-threaded application. As a result, it is more difficult to identify and verify root causes when errors occur. There involves an extra cost due to data transfers, synchronization, communication, thread creation/destruction, etc.

Acknowledgements The research presented is supported by the project NextGenSmart DC (629.002.102) funded by the Netherlands Organization for Scientific Research (NWO).

References

1. Armbrust M, Fox A, Griffith R, Joseph AD, Katz R, Konwinski A, Lee G, Patterson D, Rabkin A, Stoica I (2010) A view of cloud computing. Commun ACM, 53(4):50–58
2. Buyya R, Yeo CS, Venugopal S, Broberg J, Brandic I (2009) Cloud computing and emerging it platforms: vision, hype, and reality for delivering computing as the 5th utility. Futur Gener Comput Syst 25(6):599–616
3. Zhang Q, Cheng L, Boutaba R (2010) Cloud computing: state-of-the-art and research challenges. J Internet Serv Appl 1(1):7–18
4. Zhao H, Wang J, Liu F, Wang Q, Zhang W, Zheng Q (2018) Power-aware and performance-guaranteed virtual machine placement in the cloud. IEEE Trans Parallel Distrib Syst 29(6):1385–1400
5. Liu X-F, Zhan Z-H, Deng JD, Li Y, Gu T, Zhang J (2018) An energy efficient ant colony system for virtual machine placement in cloud computing. IEEE Trans Evol Comput 22(1):113–128
6. Li X-K, Gu C-H, Yang Z-P, Chang Y-H (2015) Virtual machine placement strategy based on discrete firefly algorithm in cloud environments. In: 2015 12th international computer conference on Wavelet active media technology and information processing (ICCWAMTIP). IEEE, pp 61–66
7. Pal SK, Rai CS, Singh AP (2012) Comparative study of firefly algorithm and particle swarm optimization for noisy non-linear optimization problems. Int J Intell Syst Appl 4(10):50
8. Jatoth C, Gangadharan GR, Buyya R (2015) Computational intelligence based qos-aware web service composition: a systematic literature review. IEEE Trans Serv Comput 10(3):475–492
9. Johnson DS (1982) The NP-completeness column: an ongoing guide. J Algorithms 3(2):182–195
10. Hartmanis J (1982) Computers and intractability: a guide to the theory of NP-completeness (michael r. garey and david s. johnson). Siam Rev 24(1):90
11. Portaluri G, Adami D, Gabbrielli A, Giordano S, Pagano M (2017) Power consumption-aware virtual machine placement in cloud data center. IEEE Trans Green Commun Netw 1(4):541–550
12. Lee S, Panigrahy R, Prabhakaran V, Ramasubramanian V, Talwar K, Uyeda L, Wieder U (2011) Validating heuristics for virtual machines consolidation. Microsoft Research, MSR-TR-2011-9, pp 1–14
13. Wang Y, Xia Y (2016) Energy optimal vm placement in the cloud. In: Proceedings of the IEEE 9th international conference on cloud computing (CLOUD). IEEE, pp 84–91

14. Sayadnavrad MH, Haghighat AT, Rahmani AM (2018) A reliable energy-aware approach for dynamic virtual machine consolidation in cloud data centers. J Supercomput 1–22
15. Shaw R, Howley E, Barrett E (2019) An energy efficient anti-correlated virtual machine placement algorithm using resource usage predictions. Simul Model Pract Theory
16. Michael R. Garey, David S. Johnson (1990) Computers and intractability: a guide to the theory of NP-completeness. W. H. Freeman & Co., New York, NY, USA
17. Vomlelova M, Vomlel J (2003) Troubleshooting: NP-hardness and solution methods. Soft Comput 7(5):357–368
18. Wu G, Tang M, Tian Y-C, Li W (2012) Energy-efficient virtual machine placement in data centers by genetic algorithm. In: Proceedings of international conference on neural information processing. Springer, pp 315–323
19. Seyed Ebrahim Dashti and Amir Masoud Rahmani (2016) Dynamic VMs placement for energy efficiency by PSO in cloud computing. J Exp Theor Artif Intell 28(2):97–112
20. Das H, Naik B, Behera HS (2019) Medical disease analysis using neuro-fuzzy with feature extraction model for classification. Inform Med Unlocked 100288
21. Das H, Naik B, Behera HS (2018) Classification of diabetes mellitus disease (dmd): a data mining (dm) approach. In: Progress in computing, analytics and networking. Springer, pp 539–549
22. Das H, Jena Ak, Nayak J, Naik B, Behera HS (2015) A novel pso based back propagation learning-mlp (pso-bp-mlp) for classification. In: Computational intelligence in data mining, vol 2 Springer, pp 461–471
23. Dey N, Ashour AS, Kalia H, Goswami R, Das H (2019) Histopathological image analysis in medical decision making
24. Sahoo AK, Mallik S, Pradhan C, Mishra BSP, Barik RK, Das H (2019) Intelligence-based health recommendation system using big data analytics. In: Big data analytics for intelligent healthcare management. Elsevier, pp 227–246
25. Rout M, Jena AK, Rout JK, Das H (2020) Teaching–learning optimization based cascaded low-complexity neural network model for exchange rates forecasting. In: Smart intelligent computing and applications. Springer, pp 635–645
26. Mohanty S, Moharana SC, Das H, Satpathy SC (2020) Qos aware group-based workload scheduling in cloud environment. In: Data engineering and communication technology. Springer, pp 953–960
27. Gharehpasha S, Masdari M, Jafarian A (2019) The placement of virtual machines under optimal conditions in cloud datacenter. Inf Technol Control 48(4):545–556
28. Grange L, Da Costa G, Stolf P (2018) Green it scheduling for data center powered with renewable energy. Futur Gener Comput Syst 86:99–120
29. Gupta MK, Amgoth T (2018) Resource-aware virtual machine placement algorithm for iaas cloud. J Supercomput 74(1):122–140
30. Li Z, Li Y, Yuan T, Chen S, Jiang S (2019) Chemical reaction optimization for virtual machine placement in cloud computing. Appl Intell 49(1):220–232
31. Gandelli A, Grimaccia F, Mussetta M, Pirinoli P, Zich RE (2007) Development and validation of different hybridization strategies between GA and PSO. In: Proceedings of the IEEE congress on evolutionary computation. IEEE, pp 2782–2787
32. Kao Y-T, Zahara E (2008) A hybrid genetic algorithm and particle swarm optimization for multimodal functions. Appl Soft Comput 8(2):849–857
33. Esmin AAA, Lambert-Torres G, Alvarenga GB (2006) Hybrid evolutionary algorithm based on PSO and GA mutation. In: Proceedings of the sixth international conference on hybrid intelligent systems. IEEE, pp 57–57
34. Shi XH, Liang YC, Lee HP, Lu C, Wang LM (2005) An improved ga and a novel PSO-GA-based hybrid algorithm. Inf Process Lett 93(5):255–261
35. GáLvez A, Iglesias AS (2013) A new iterative mutually coupled hybrid GA-PSO approach for curve fitting in manufacturing. Appl Soft Comput 13(3):1491–1504
36. Shi XH, Lu YH, Zhou CG, Lee HP, Lin WZ, Liang YC (2003) Hybrid evolutionary algorithms based on pso and ga. In: Proceedings of the congress on evolutionary computation, CEC'03., vol 4. IEEE, pp 2393–2399

37. Kennedy J (1997) The particle swarm: social adaptation of knowledge. In: Proceedings of the IEEE international conference on evolutionary computation. IEEE, pp 303–308
38. Del Valle Y, Venayagamoorthy GK, Mohagheghi S, Hernandez J-C, Harley RG (2008) Particle swarm optimization: basic concepts, variants and applications in power systems. IEEE Trans Evol Comput 12(2):171–195
39. Reeves CR (1993) Using genetic algorithms with small populations. In: Proceedings of the Fifth international conference on genetic algorithms. Morgan Kaufmann, pp 92–99
40. Beloglazov A, Abawajy J, Buyya R (2012) Energy-aware resource allocation heuristics for efficient management of data centers for cloud computing. Futur Gener Comput Syst 28(5):755–768
41. Carlisle A, Dozier G (2001) An off-the-shelf pso [c/cd]. In: Proceedings of the workshop on particle swarm optimization. Indianapolis
42. Schutte JF (2001) Particle swarms in sizing and global optimization. Master's Thesis, University of Pretoria, Department of Mechanical and Aeronautical Engineering
43. Felician A, How to parallelize a sequential program
44. Amdahl GM (1967) Validity of the single processor approach to achieving large scale computing capabilities. In: Proceedings of the April 18-20, 1967, spring joint computer conference. ACM, pp 483–485
45. Calheiros RN, Ranjan R, Beloglazov A, De Rose CAF, Buyya R (2011) Cloudsim: a toolkit for modeling and simulation of cloud computing environments and evaluation of resource provisioning algorithms. Softw: Pract Exp, 41(1):23–50
46. Wang S, Liu Z, Zheng Z, Sun Q, Yang F (2013) Particle swarm optimization for energy-aware virtual machine placement optimization in virtualized data centers. In: 2013 international conference on parallel and distributed systems. IEEE, pp 102–109
47. Premalatha K, Natarajan AM (2009) Hybrid pso and ga for global maximization. Int J Open Probl Compt Math 2(4):597–608
48. Gavvala SK, Jatoth C, Gangadharan GR, Buyya R (2019) Qos-aware cloud service composition using eagle strategy. Futur Gener Comput Syst 90:273–290
49. Virtual library of simulation experiements: test functions and datasets. https://www.sfu.ca/~ssurjano/index.html
50. Benchmarkfcns toolbox. http://benchmarkfcns.xyz/fcns
51. Reddy VD, Gangadharan GR, Rao GSVRK (2017) Energy-aware virtual machine allocation and selection in cloud data centers. Soft Comput 1–16
52. Yapıcı H, Çetinkaya N (2017) An improved particle swarm optimization algorithm using eagle strategy for power loss minimization. Math Probl Eng

Chapter 5
Root-Cause Analysis Using Ensemble Model for Intelligent Decision-Making

Sheba Selvam, Blessy Selvam, and J. Naveen

Abstract The increase in social media and social media applications made a way for people to opt for online based decision-making. Before this progression, people used to discuss with friends, colleagues and relatives to purchase some products. Currently the social media has taken over the decision-making through online reviews. As a result, large organizations are dealing with such reviews to provide huge benefits. This enormous information available on the web is looked-for by both the organizations and the customers. This has also made decision weariness owing to the devastating number of opinions on a single product. To overcome this issue, root-cause analysis model is proposed for effective decision-making. The root-cause analysis model in turn consists of multiple models namely; Aspect categorization Ontology for aspect extraction, Prediction-based word embedding model, Variegated Ensemble-based Weighted Voting (VEWV) Model for prediction is used to reduce the computational complexity levels and error levels, and Ontology Reinforcement for frequent updates in the ontology system. The experiments were examined with the five-product dataset and compared with recent works. The proposed model shows enhanced performance in terms of precision, recall and F-Measure.

Keywords Aspect extraction · Root cause analysis · Aspect categorization ontology · Word embedding · Ensemble modeling · Random forest

---

S. Selvam (✉)
Department of Computer Science and Engineering, RajaRajeswari College of Engineering, Bangalore 560074, Karnataka, India
e-mail: shebanaveen@gmail.com

B. Selvam
Department of Computer Science and Engineering, Anna University (BIT Campus), Trichy 620024, Tamil Nadu, India
e-mail: blessyselvam19@gmail.com

J. Naveen
Department of Computer Science and Engineering, CHRIST (Deemed to Be University), Bangalore 560074, Karnataka, India
e-mail: naveen.j@christuniversity.in

J. K. Rout et al. (eds.), *Machine Learning for Intelligent Decision Science*, Algorithms for Intelligent Systems, https://doi.org/10.1007/978-981-15-3689-2_5

1 Introduction

There is a trend of increasing use of social media applications due to the rise in information availability. Earlier, peers and relatives were consulted before a purchase decision was to be made. Today, online reviews of products provide insight to the customers who influence their decision-making. Both, consumers and organizations have benefited from the web which has become a vast source of information [1]. Reviews and opinions available on the web correspond to customer reviews [2]. The real sentiments of consumers were represented through this information. This is analyzed by organizations to obtain information which is really a reliable source of information, and helps in decision-making. This information is also used to make purchase decisions by users, as they are mostly unbiased [3]. It is challenging to analyze this information.

The reviews are also sought after in the electronic media rather than in-person. A lexical parser and a module to identify the sentiment or polarity associated with the text is required to perform analysis on input which is in text form. Extracting valuable information from blocks of text is known as text mining. From a business point of view, information extracted from mining corresponds to business insights. The general elements involved in a text mining process include pre-processing of text, transformation of textual data to numerical data, feature selection and performing data mining on the generated content. Further analysis is carried out on insights obtained from the mining process to aid in decision-making process. This procedure of determining the sentiments associated with a text is known as Opinion Mining. The process of identifying the major components of a text is known as Aspect Mining. Aspect corresponds to an opinion target. The target may be a product or the feature of a product. Aspects generally refer to product features since data retrievals are dependent upon products in real time [4]. Considering the statement 'The picture quality is extraordinary', picture quality is the aspect and it exhibits a positive sentiment. Sentiments, either positive or negative, corresponding to the major cause, while the aspects correspond to sub-causes. Aspect extraction can be categorized into two major types [5] implicit aspects and explicit aspects [6]. Explicit aspects can be directly derived from the text, while implicit aspects have to be determined. Explicit aspects are clear and are identified directly from the text, while implicit aspects are hidden and an analysis is necessary to identify the actual aspect. For example, in the sentence, "the screen resolution is great and the phone is reliable for outdoor trips even without power banks", screen resolution is the explicit aspect, while battery performance is the implicit aspect. Organizations use this information to determine the positives and negatives about their products to further aid its enhancement. Performing manual analysis is impossible due to the vast amount of data that is available. This mandates the need for technologies to determine root causes that help in decision-making [1]. Root-Cause analysis is a model developed to provide aggregated and accurate views of products to avoid decision fatigue for users and ensure effective decision-making for organizations. Root-Cause Analysis (RCA) is the process of determining the major causes from a text corpus. Due to the vast amount of information available online, Root-Cause analysis has become the primary demand for organizations. Rather than being a single model, root-cause analysis is a composition of several multiple components, namely; aspect extraction or opinion target extraction and sentiment identification or opinion mining [7]. The process of identifying the overall stand-point of the reviewer on a certain product is called Sentiment Analysis [8]. Individuals are assisted to make effective decisions and organizations are facilitated to identify the strengths and weaknesses of their products [9]. Effective Natural Language Processing (NLP) techniques are required to uncover the root causes in the information available online which is in textual form. The major components of NLP are Aspect extraction and sentiment analysis [10]. Techniques like ensemble modeling and deep learning have been popularized for such analysis. Combining several algorithms to perform predictions is known as Ensemble modeling [11, 12]. Since an ensemble model combines multiple machine learning algorithms, it provides better prediction compared to other models that utilize a single model for prediction. It is also because of its heterogeneous nature.

The transition of web from information supplier to information receiver has reshaped the outlook of organizations and customers. Users have the liberty to voice their opinion due to the increased availability of information. As a result, a plethora of information is published online. Customers may be helped by providing a method to sift through this abundance of information. This is called Root-Cause analysis, which is the major scope of the proposed work.

2 Literature Review

With the growing use of the web and its related components, immense amount of opinion on a singular product leads to decision fatigue which is a bane of this plethora of information. Decision-making process in organizations is influenced by the information on social media. More and more people rely on other customer's reviews and comments prior to making purchasing decisions [13]. A recent technique known as root-cause identification gained importance. The significant components of root-cause identification architecture are sentiment analysis or opinion mining and aspect identification which had been investigated by the research community for some time. In this section, the prevalent and most noteworthy research in the domain of opinion mining are discussed. A literature review was performed in the four sections namely; unsupervised approaches, semi-supervised approaches, supervised approaches and rule-based approaches. Machine learning models form the basis of the initial sections while linguistic rule model is focused on in the final section.

2.1 Unsupervised Approaches

Unsupervised models are developed for the prediction process without human intervention or supervision. The initial study on aspect extraction was performed by Hu et al. [4]. This is a rule-based model which operates by identifying nouns, considering them as potential aspects. Association rules for generating candidate aspects is also utilized. An enhancement to this model was proposed by Popescu and Etzioni [14], Popescu et al. [15]. A web-oriented similarity model which utilized frequency-based pruning techniques was proposed by Li et al. [16]. For the aspect term and the entity, the number of hits generated from a search engine is determined. It is then used for identifying similarity levels for the final results. Other aspect mining models based on frequency include works by Eirinaki et al. [17] and a model that uses tourism-based reviews proposed by Marrese-Taylor et al. [18]. Other approaches are the models proposed by [12, 13] and Lin et al. [19], that uses bootstrapping-based techniques.

Identification of a subset of the available attributes is called feature selection. For improved results, feature selection helps in decreasing data that helps in faster processing. A K-Means clustering based approach was proposed by Abualigah and Khader [20]. This work ensures effective prediction by examining the quality of selected attributes. The hybrid particle swarm optimization (H-FSPSOTC) algorithm is a hybrid multi-model-based algorithm. When compared with other models, this model shows better outcomes. An opinion mining model that operating on entity-level was proposed by Khan et al. [21]. This technique demonstrates high recall and low precision utilizing lexicon for analysis. Better recall levels were provided by the automated model that performs opinion identification on twitter data. Using a classifier model, the polarity is identified. Based on opinions, polarity is allocated to the tweets.

2.2 Semi-supervised Approaches

Both supervised learning, i.e., rule building by training data and user's interaction, is utilized by semi-supervised models. A bootstrapping-based model for aspect extraction and opinion mining was provided by Hai et al. [22]. This model is language specific and is based on Chinese reviews. A model using Latent Dirichlet Association (LDA) for aspect extraction was proposed by Ma et al. [23]. A word alignment based model which determines aspects based on word positions is another model that uses semi-supervised approaches as proposed by Liu et al. [24]. Other models include an association-based aspect mining model by Yan et al. [25] and a review-based aspect mining model by Samha et al. [26]. An architecture for the general operational nature of the semi-supervised aspect extraction model was proposed by Wang et al. [27]. A Semi-supervised alignment model was proposed by Liu et al. [24], where opinions and their relationships are identified. Wu et al. [28] proposed a combined unsupervised model for aspect term and opinion extraction. This is a hybrid model combining rules and the machine learning process for effective identification of aspects.

2.3 Supervised Approaches

Labeled data is used by machine learning models for the training process in the supervised approach. Cruz et al. [29] proposed a taxonomy-based model that develops aspect categorizations and work on specific data. Other customer review based models include works by Chen et al. [30], Yang and Cardie [31] and Huang et al. [5]. Heuristic and Metaheuristic models are also currently used in this domain with the rise in data used for processing. A binary classifier model that uses heuristics was proposed by Li et al. [32, 33]. The aim of this model is to map opinion targets. Poria et al. [34] proposed a neural network-based model which utilizes a 7-layer convolutional neural network for deep learning. In [35], analysis was performed on medical data for disease classification using Linguistic Neuro-Fuzzy model with feature extraction. In [36] various machine learning classification algorithms are experimented in 14 different biomedical datasets, and derived to a conclusion that classification accuracy depends on the nature of the dataset. Classification algorithms J48 and Naïve Bayes [37] were used on the medical dataset of patients to predict diabetes.

Jeyapriya and Selvi [38] proposed a supervised model to extract implicit aspects from reviews. Based on word co-occurrence and word frequency levels., implicit aspects were determined and scored. A weighted aspect extraction model that also identifies the polarity levels based on user's interest was proposed by Liu et al. [24]. Li et al. [32] proposed a pattern-based and polarity based model which operates by bootstrapping extracted aspects. Similarity identification follows aspect extraction. He et al. [39] proposed a neural attention method used for aspect extraction on review data. In this work, an ABAE (Attention-based Aspect Extraction) method is proposed that captures word co-occurrence patterns explicitly. A word embedding mechanism to extract the feature vector is used by this attention-based aspect extraction and addressed the data sparsity issue. Paredes et al. [40] proposed a Convolutional Neural Network (CNN) that was utilized for aspect extraction and sentiment analysis in Spanish product reviews. A deep CNN was proposed to build a classifier for identification of sentiments on Twitter data. Improved results were attained not considering the temporal features.

Ruder et al. [41] proposed a deep learning approach for aspect extraction and opinion mining using CNN. A deep convolution neural network across multilingual domains on laptop, hotel, phone and camera datasets was proposed in this work. Hu et al. [4] proposed a deep neural network where the domain and linguistic features were extracted. A plethora of reviews were considered and classified using deep neural network. Movie and hotel reviews were worked on by the deep neural network. Deep learning is now a coveted model for sentiment analysis. Intelligent decision-making in healthcare [42–44] has been focused in real-time data for predicting diseases using deep learning models [25, 45–47]. Poria et al. [34] proposed a deep convolutional network-based aspect extraction model. Zhang et al. [48] proposed a deep learning-based model which utilizes both CNN and RNN for sentiment analysis. The advantage of this model that makes it effective is that storage of long-term dependencies and temporal features can efficiently be handled by RNN while, identification of potential features can be carried out by CNN. However, due to the substantial use of deep learning methods, the model is computationally complex. Neural network models, due to their complexity in handling features, were further enhanced to perform multiple processes. They provide substantial advantages to applications involving NLP. Texts in languages other than English can be processed using deep learning techniques. Zheng et al. [49] proposed a Chinese review processing system for POS tagging and word segmentation.

Collobert and Weston [50] proposed a unified network architecture model that is a deep learning model which provides a single architecture for the identification of POS tags, named entities and semantically similar elements in the text making it a multitasking architecture. Majumder et al. [51] proposed a deep learning model for personality detection. Multi-layered network along with pooling and convolutional layers are utilized in this model for effective predictions. Asghar et al. [52] proposed a heuristic-based aspect extraction model. This was developed by extending the heuristic patterns as an integrated framework for enhanced aspect extraction. A hybrid sentiment classification module is included in the model along with a summary generator for enhanced performance. Akhtar et al. [53] proposed a two-step aspect extraction model which was an alternate heuristic based model that utilizes PSO for sentiment analysis and a broadened rule set for aspect extraction. The submitted models showed good performance levels but were not real-time compliant due to their high computational complexity. PSO based evolutionary multilayer perceptron [54] is proposed which is intended for classification task in data mining.

2.4 Rule-Based Approaches

Syntactic patterns are incorporated in rule-based models to develop conditions for prediction. The earliest, most significant contribution was made by Qiu et al. [55] in the domain of aspect extraction. Double Propagation (DP) is a method proposed in this work is noteworthy in this domain. A rule-based technique based on bootstrapping is proposed in this work enabling efficient identification of aspects from a given text. Hu and Liu [56] in another previous work proposed a summarization-based model which enables identification of primary components in a text. Popescu and Etzioni [14], Popescu et al. [15] further enhanced this model to extract product features and their corresponding opinions from product reviews. Scaffidi et al. [57] proposed another similar product feature extraction model that utilizes a language model to identify product features.

Poria et al. [34] submitted a deep learning model for aspect identification that provides predictions by employing convolution neural networks to map language intricacies. Rana and Cheah [6] proposed a work which examined the effectiveness of aspect extraction in sentiment analysis process. Rana and Cheah [58] produced a follow-up work, a twofold rule-based model for aspect extraction. Aspects and opinions are identified in the initial phase. This process is independent of the domain. Domain-dependent details are extracted in the second phase. A good result is ensured by the combination of both these methods. Liu et al. [8] proposed a similarity-based model which utilizes association-based recommendations. Kang et al. [59] proposed a frequency-based extraction model. Kang and Zhou [60] also submitted a rule-based aspect extraction model. Identification of both objective and subjective features from the text is done in this work. Review-specific patterns provide objective features while broadening the DP model provides subjective reviews.

A spectator model was proposed by Hai et al. [61] which performs automatic identification and rating of aspects in reviews. The dependency path syntactic between different words is pared in the model. A syntactic dependency tree is generated by the model after performing opinion mining initially. Identifying the strongest opinion pairs is the next step in the process. The opinion mining process utilizes WordNet and SentiWordnet. A model was proposed by Wang et al. [62] to identify the relevance for product aspects. Identification of aspect and opinion pairs is done in the model. Based on Intrinsic and Extrinsic domain relevance, scoring is carried out. Domain relevance values were used to identify aspects associated with opinions. Liu et al. [8] proposed an aspect extraction model that utilizes double propagation-based recommendation for product reviews. To improve the syntactic rule-based model, a two-level recommendation is proposed in the work wherein the first level, semantically similar aspects are obtained followed by the extraction of associated aspects and then combined for ranking. A crucial role is played by ontologies in the identification of related information which enables efficient and reliable information retrieval. Munir and Anjum [63] proposed a summary of the models based on ontology-based information retrieval. The models are discussed in three parts. Ontology-based query formulation is presented in the first part. Information linking approach based on ontology is presented in the second. Query refinement strategies are revealed in the final section. Hybrid forecasting model [64] is proposed to predict exchanges of currency using Teaching-Learning-based optimization (TLBO). Sun et al. [65] proposed a model that works with electronic word-of-mouth information for products. Online customer reviews form the base data. This model performs fine-grained sentiment analysis. Lau et al. [66] provided a Latent Dirichlet Allocation (LDA) based model for ontology construction based on product data. The taxonomic and non-taxonomic relationship between aspects is identified by the model. Amplayo and Song [11] proposed a similar model using LDA. Other models include analysis-based studies by Konys [67].

3 Proposed Work

Social media is ruling the entire world for the past two decades. The progression of WWW made all information available right in the hands of an individual. This enormous data in the social media made a way for decision-making in many organizations, even customers purchase products deciding on the reviews, decides a restaurant for food based on the online reviews. The aim of the proposed work is in identifying the root causes for effective decision-making by aspect-based opinion mining. Identifying the root cause involves textual analysis along with human cognition. The proposed work is given in Fig. 1 and involves:

- Pre-processing of reviews
- Aspect categorization Ontology for aspect extraction
- Prediction-based word embedding model
- Variegated Ensemble-based Weighted Voting (VEWV) Model for prediction
- Aspect Ranking and Ontology Reinforcement.

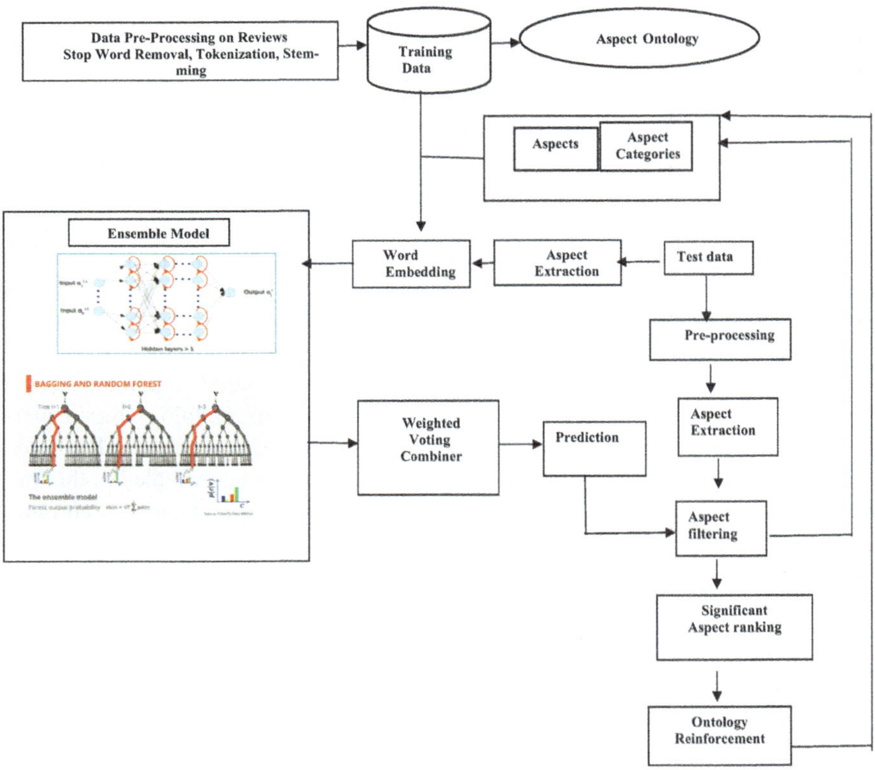

Fig. 1 Proposed ensemble model for intelligent decision making

3.1 Pre-processing of Reviews

The Preliminary step of the proposed work starts with Pre-processing of the reviews. Since these reviews are from the social media it has to be pre-processed before aspect extraction. In the reviews, text, punctuations and stop words are present. Text input is tokenized in the first phase. Tokenization is the process of breaking down text into many components, each having a single meaningful word, called a token. Stop words are filtered in the preceding text. Content that does not contribute to add any meaning to the text are termed as stop words. They act as filler words. Further processes include handling of negations, followed by stemming. The process of determining core words is called stemming. Tokens are classified into corresponding elements using data in this stage. A sentiment class is prepared by aggregating the resultant content in the document.

3.2 Aspect Categorization Ontology

The proposed work was initiated by constructing a domain ontology for aspects and aspect categories followed by ontology update modules for maintenance. Identifying explicit aspects are easy. There is a need for ontology to extract the implicit aspects. Using Parts of Speech (POS) components on the training data, aspects are extracted. POS tags are examined and nouns, noun phrases and multiword aspects are mined using well-defined rules [55]. A model creation phase was done to derive important entities for training data.

Significant terms were extracted from the input information as aspects to build domain-based ontology. The significant terms identified from the set of reviews of a product or service are known as entity. The entity forms the highest-level categorization in an ontology. The aspect category falls in-between the entity and the aspect. For example, in the sentence "the mobile phones' battery is good, but it is costly," the entity is mobile phone. The entity has aspects: battery and costly, and aspect categories: phone and prize.

Aspect identification is very important phase for ontology construction. Aspect links to an opinion which describes the concept relating to an entity. From the input training information, the aspects are extracted and grouped into sections. These are known as aspect categories. The aspect categories highly rely on the domains and hence manually created. A metadata table is maintained to identify Occurrence Frequency and Distribution of Frequency values. Two tables namely aspect metadata and aspect category metadata are maintained. The aspect metadata accumulates the value of Aspect, Occurrence Frequency, Aspect Weight, Positive Frequency Distribution and Negative Frequency Distribution. Occurrence Frequency values the number of times the aspects and aspect categories appeared in the training data. The importance of the relating aspect is signified in the Aspect Weight and Aspect Category Weight by the ratio between the occurrence frequency and total number of reviews in the training data. The total positive reviews for an aspect or aspect category fills the Positive Frequency Distribution and the total negative reviews correspond to the Negative Frequency Distribution.

The entity is the root element in aspect ontology creation. The other elements represent aspect categories in the tree structure. Some of the aspects may relate to multiple aspect categories, which are added to multiple groups to represent the aspect categories. Finally, the Aspect Ontology system components are Aspect, Category, Occurrence Frequency, Positive Frequency Distribution, Negative Frequency Distribution and Time Stamp. Time stamp remains the same for aspects during the ontology creation later it varies based on its occurrence.

3.2.1 Ontology Update Using LRU Operation

The occurrence of an aspect plays an important role in prediction. In some cases, some aspect would have been prioritized in the past, but its significance minimal in due course of time. Numerous new-fangled aspects would have occurred recently, which needs to dominate the prediction model in the ontology system. As a result, usage of frequency occurrence leads to old information. To overcome this issue, the proposed model works by updating the aspects which reflects in the Time Stamp. Based on the timestamp, the recent aspects are pushed to the top of the list in the ontology system. This model represents the Least Recently Used (LRU) operation. The list is continuously updated to hold recent aspects in the ontology list. By this LRU operation, the old and stale information are removed from the list, empowering efficient ontology list maintenance.

3.3 Prediction-Based Word Embedding Model

As the reviews were textual information, it could not be used directly to train the machine learning models. So, for effective processing there was a need to convert them to numerical vectors. This process of converting textual information into numerical vectors is known as word embedding model. In the proposed work, Skip-gram model and Continuous Bag of Words (CBOW) is used for the word embedding task as it reduces the size of the data dimension. This CBOW is a prediction-based model. This numerical vector captures the context of the word, its relationship with other words representing the content, as well as its syntactic and semantic similarities.

3.4 *Variegated Ensemble-Based Weighted Voting (VEWV) Model for Prediction*

This work utilizes a Variegated Ensemble-based Weighted Voting (VEWV) for necessary prediction. The word embedding of the training data is carried out by passing the training data to the word embedding module for conversion. The word embedding module converts the verbal representation of the training data to a numerical vector format. The latter is passed to the ensemble model for developing the training model. The vectors that are in numerical format, which was created previously, were passed to the ensemble model.

3.4.1 Variegated Ensemble Model Creation

A training model is created using the proposed Variegated Ensemble-based Weighted Voting (VEWV) model. Textual data is the base for predicting the level of sentiments related to the reviews. Data being textual contains several obscurities incorporated with it. Even though the data is converted to numerical formats, the obscure details of text still exist in the numerical vectors. Especially human generated data does not really assure with the act of the language. This entails the need for multiple models, as a single model cannot effectively record all the obscurity information on the text. The proposed Variegated Ensemble model is created with two models namely; Random forest classifier and LSTM (Long short-term memory). A weighted voting combiner is used to combine the independent predictions given by each model in order to form a single prediction set. The single prediction set is the combination of all predictions provided by the ensemble models.

3.4.2 Sentiment Identification

The feed forward neural network and the recurrent neural network were commonly used in Root-Cause analysis due to its performance. But it forgets the previously read information because of which the long-term dependencies are hard to find. It is a big issue, as the training process gets harder as it cannot guess the next data to be read. Also, it is difficult to deal with error information; weights are missing which is a drawback for RNN. Hence in the proposed work LSTM is used.

Long Short-Term Memory (LSTM) networks are a revised form of recurrent neural networks, which remembers the past data in memory easily [34]. The LSTM overcomes the problem of vanishing gradient; it trains the model using back propagation. LSTM classifies process and predicts data easily as it solves the vanishing gradient problem.

LSTM (Long short-term memory) influences the data storage and the inflow and outflow based on the training process, not only that but it controls the overall storage of weights, keeps a hold on previous data, take out unwanted data and transforms data to the next step. The training period is reduced significantly; as it adheres to erroneous information for a longer period to keep gradients relatively steep [61]. However, LSTM tries to remember only the most significant parts of the state and new data.

In an LSTM network three gates are present

1. **Input gate**—the input gate finds the value from input that needs to be used for modifying the memory. Sigmoid function sets a range between 0 and 1. The values near to 1 are kept and those near to 0 are removed. In the same way the tanh function provides weightage to pass values. The significant level of the values is based on the values ranging between −1 to 1.
2. **Forget gate**—the values from the previous state and the current input state are passed to the sigmoid function where the values appear between 0 and 1. The values near to 0 are removed and the values near to 1 are kept.
3. **Output gate**—the value from input that needs to be used for modifying the memory includes (previous state and the current input state). This information is passed to sigmoid function which accepts values near to 1 and removes values those are near to 0 are. The same information is passed to tanh gives weightage to the values which are passed and outputs values between −1 and 1. The values near to 1 are considered relevant and those near to −1 are considered irrelevant. The product of sigmoid, tanh along with sum of output of forget gate is obtained as cell state. The product of tanh cell state and sigmoid of previous state and current state input are obtained. The resultant product is the new hidden state.

$$i_t = \sigma(w_i[h_{t-1}, x_t] + b_i) \tag{1}$$

$$f_t = \sigma(w_i[h_{t-1}, x_t] + b_i) \tag{2}$$

$$o_t = \sigma(w_i[h_{t-1}, x_t] + b_i) \tag{3}$$

In the above Eqs. (1, 2 and 3), i_t, f_t and o_t represents input, forget and output gates, w_y represents weight for respective gate, h_{t-1} output of the previous block, x_t represents current input timestamp, b_y represents bias for the respective gate and σ represents sigmoid function.

It takes as input the previous prediction, previous state and the new information. The previous prediction and new information are fed separately into four different neural networks. LSTM combines a neural network with the new and previous information, combined with the previous state as shown in Fig. 2. A training set is needed to train these neural networks, which in turn makes up the LSTM cell.

Boosting is an ensemble model that makes a base learner to reduce the bias in the prediction methodology by performing iterative training and thereby including the error levels [68]. The proposed work uses **LSTM**. However, this model tends to induce overfitting in the prediction mechanism. However, the calculations often get

Fig. 2 Work flow of LSTM model

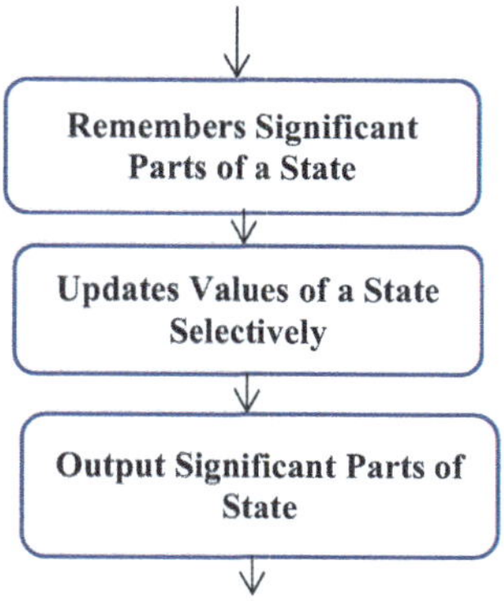

very complex with increase in attributes. In-order to compensate for the pitfall and to handle the overfitting problem the proposed work included ensemble Random forest based boosting prediction model [69].

3.4.3 Random Forest Model

Random Forest Classifier is ensemble algorithm [19]. Ensemble algorithm combines two or more models of homogeneous or heterogeneous type for prediction classification. It calculates the relevance score for each feature, which is computed automatically in the training phase. Then the relevance score is scaled down, such that the sum of the scores is equal to 1, with the help of this score the significant aspects are obtained and thereby eliminating the least important aspects for building the model [70]. Such elimination sounds the F-measure of the model.

Random forest reveals the importance of features in the prediction [71]. Random forests create decision trees on data samples that are selected randomly, obtain the prediction from each one of the trees and conducts voting on these predictions in order to obtain the best prediction.

The Random forest works as

It works as follows:

- Random samples are selected from the dataset.
- For each sample decision tree is constructed and independent predictions from each tree is obtained.
- Voting is made for each subset of prediction.
- The prediction that has been voted the most is used for obtaining a final prediction.

The proposed method works well, as a single decision tree may be prone to noise, but collection of decision trees reduces noise and hence promote for accurate results and there by canceling out the biases [59]. A fewer quantity of features reduces the training time. Therefore, it will not suffer from the issue of over fitting. The ensemble model is a combination of two techniques, which reveals better results.

The model training is carried out by progressing the numerical vectors to the two models. The entire training data is progressed to the two models, where the training occurs independently.

The ensemble model performs prediction as the data to be predicted are passed to it; here a prediction set is produced by each of the models. Hence, leading to multiple predictions at every instance. This phase combines the predictions obtained by each model into a single prediction set by using the weighted voting mechanism [57]. The prediction weights are based on the model's performance on the training data. Weights are assigned in terms of the level of probability prediction. Hence sum of weights will be equal to 1. This is given by Eq. (4)

$$v_1 + v_2 + v_3 = 1 \quad (4)$$

where v_1, v_2 and v_3 are the weights associated for each prediction model in the ensemble. The process of identifying the final prediction is given by Eqs. (5) and (6)

$$v_1 + v_2 + v_3 = 1 \quad (5)$$

$$\text{Final Prediction} = \begin{cases} 1 \text{ if Prediction} \geq 0.7 \\ 0 \text{ Otherwise} \end{cases} \quad (6)$$

where R_1, R_2 and R_3 are the predictions given by the models and v_1, v_2 and v_3 are their corresponding weights.

3.5 Aspect Ranking and Ontology Reinforcement

The significant aspect identification in the text is essential for identifying the root causes. Predictions obtained from VEWV model provides the level of sentiments correlated in the review text considered for analysis. The significant terms are extracted in this phase; insignificant terms are filtered based on the ontology. The ontology created already holds the significant terms up to date and once the reviews are posted newly again the significant terms are identified and this process continues. At each point of time the ontology is updated. The significant and filtered terms are integrated with the corresponding sentiments, thereby providing ranked root-causes.

3.5.1 Significant Term Extraction

Every term plays an important role in the text. Retrieving highly significant terms helps in identifying the major components of the text. The subjective terms carry no meaning, whereas the objective terms hold the meaning. So, in order to sought out the meaningful terms which holds the sentiment is mandatory for Root-Cause identification. The significant term carries meaning and also has an impact on text. This can be done by identifying the Term Frequency and the Inverted Document Frequency of the terms in text [42].

Term Frequency/Inverted Document Frequency (TF-IDF) provides the tokens significant levels in a document by comparing it with large document collections [72]. The value determined directly corresponds to the significance of the token in the given text. Further, this model also helps determine common words, hence eliminating them to provide a clean information context. The process of determining TF-IDF is given by Eq. (7),

$$\mathrm{TF}(t, d, D) = \mathrm{tf}(t, d) \times \mathrm{idf}(t, D) \tag{7}$$

3.5.2 Filtering Aspects Using Ontology

Aspect filtering is carried out by selecting the highly significant terms. Terms with higher TF-IDF values than mentioned threshold levels are accounted as significant terms. The threshold level is set to 0.08. The terms with term weight greater than the above threshold values are considered as significant terms [61]. The significant aspects considered are filtered and cross-indexed with the aspect ontology to find the aspects with higher latest levels. The highly valued terms are appended to the aspect ontology.

3.5.3 Sentiment Integration and Significance-Based Ranking

The aspects are filtered, latter is integrated with their corresponding sentiment obtained from VEWV model and perform aspect segregation. The segregation of aspects divides the aspects based on the level of sentiment [52]. This process helps in identifying the implicit aspects hidden in the reviews. The Aspect category is identified and listed based on the aspects pertaining to it. The aspect which appears again and again is filtered in order to avoid repeated aspects occupying unwanted space [17]. The ranking of aspects is done according to their Significance level.

3.5.4 Ontology Reinforcement

The existing aspect timestamps are kept up to date. positive feedback leads to reinforcement. When some of the aspects were unidentified by the model goes through the aspect extraction stage for identifying them [67]. Once the aspect that has been missed out is not available in the ontology, the POS analysis of the necessary text is carried out in order to confirm that the term is really an aspect, if confirmed; it is added as a new component to the ontology.

4 Results and Discussion

The proposed VEWV model is executed in the Spark architecture and implemented in PySpark. A benchmark customer review dataset has been used in the proposed method, which (Rana and Cheah [6]) comprises reviews from customers corresponding to five distinct products. The aspects and polarities in the datasets were annotated. Categorizations of aspects were manually defined. The performance comparison was considered with three state-of-the-art models, presented by Popescu and Etzioni [14] and models CNN+LP presented by Poria et al. [34] and TF-RBM model presented by Rana and Cheah [58]. The standard evaluation metrics for any opinion mining system is considered in this work they are precision, recall and F-measure. The proposed VEWV model results are shown in Figs. (3, 4, 5, 6, 7 and 8). The proposed model could be observed to exhibit high Recall, Precision levels, exhibiting the efficiency of the proposed VEWV model.

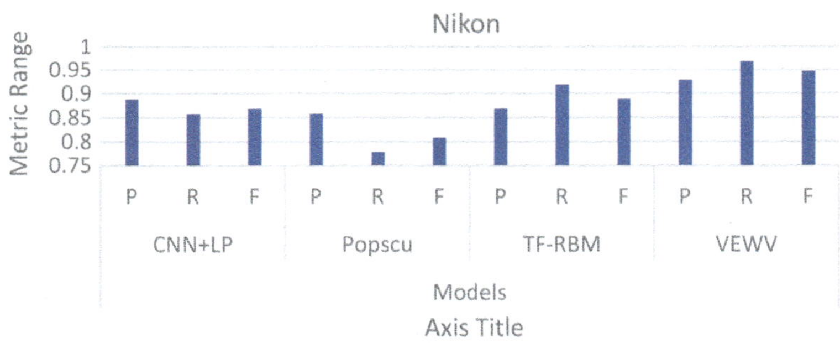

Fig. 3 Comparison chart of models on Nikon Data

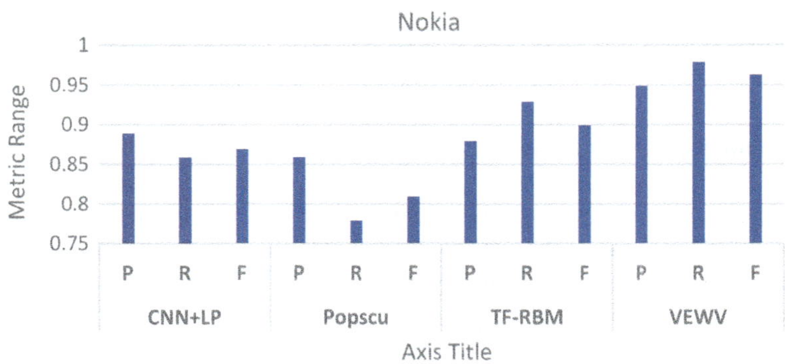

Fig. 4 Comparison chart of models on Nokia Data

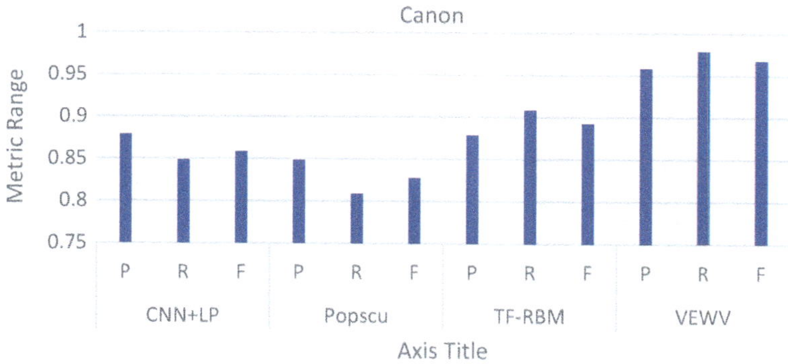

Fig. 5 Comparison chart of models on Canon Data

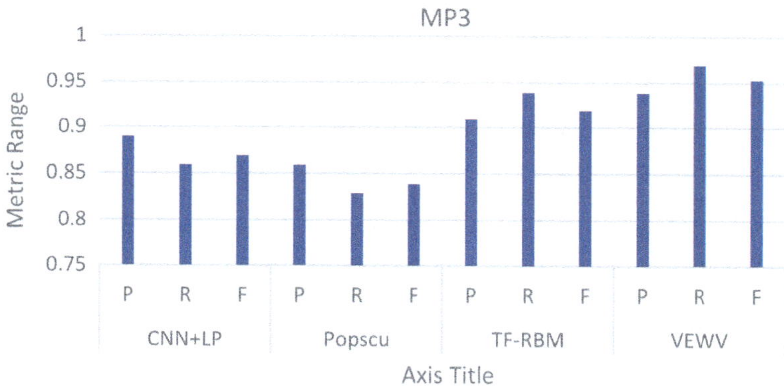

Fig. 6 Comparison chart of models on MP3 Data

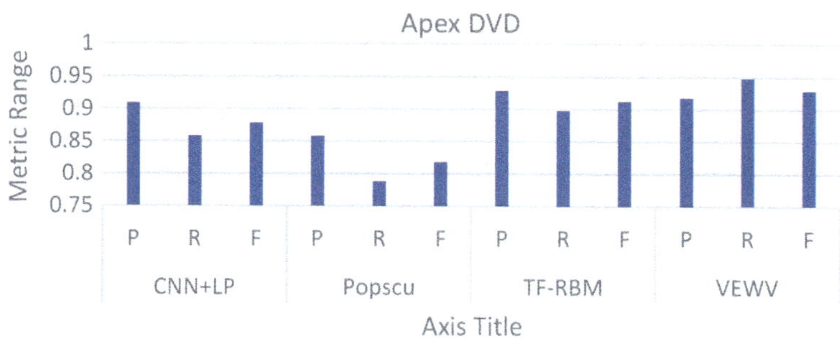

Fig. 7 Comparison chart of models on Apex DVD Data

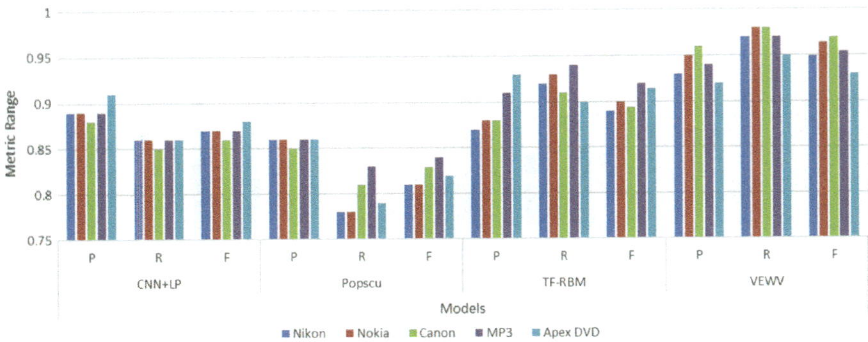

Fig. 8 Comparison chart of models on different datasets

The enhanced performance levels are achieved because of these important reasons. First, due to the aspect extraction based on ontology. In the proposed work, the ontologies are extracted based on the semantic correlation with the input query, instead of techniques like content matching. This leads to retrieval of more relevant and context-sensitive results of the proposed model, which in turn enhances the performance level of the proposed VEWV model.

The proposed model performance compared with the existing models CNN+LP, TF_RBM and the proposed Popscu models are shown in Figs. (3, 4, 5, 6, 7 and 8). It is revealed that the proposed VEWV model presents elevation in each of the metrics on all the datasets and thereby promoting greater efficiency and performance levels of the proposed work.

An average of the obtained performance metrics is shown in Table 1. The best performances are highlighted in the tabulation.

It could be observed that the proposed VEWV model exhibits improved performances of 5–9% in terms of precision, 5–18% in terms of recall and 5–13% in terms of F-Measure when compared to other models. The improved performance levels are attributed to ensemble-based model, enabling low overfitting and high performance in the models. The results in Table 1 clearly highlight the performance of the proposed model for solving real-world problems.

Table 1 Average performance metrics

	Precision	Recall	F1-Score
CNN+LP	0.892	0.858	0.87
Popscu	0.858	0.798	0.8218
TF-RBM	0.894	0.92	0.904
VEWV	0.94	0.97	0.9532

5 Conclusion

The proposed Variegated Ensemble-based Weighted Voting (VEWV) for Root-Cause Analysis model carry out three key stages namely; aspect extraction module, sentiment prediction and ranking module. For building aspect ontology and maintaining the aspect categories the training data is used. The ontology helps in ranking the latest aspects effectively by identifying the latest aspects. The aspect extraction module identifies the significant aspects and the sentiment analysis module identifies the prediction using the VEWV architecture. The integration of aspects with the related predictions is made to produce updated and final root causes.

The significant advantage of the proposed model is that the (VEWV) architecture makes use of multiple models for predictions. The significance-based ranking module enables effective ranking of the root-causes, hence the major root-causes are automatically highlighted. Although making use of two distinct models enlarges the complexity of the model, the proposed model is executed using PySpark to qualify parallel processing. Hence the actual time requirements are reduced. This qualifies the architecture to provide results in real time. The stale information in data is reduced by using timestamp-based processing, thereby validating valuable predictions.

The proposed model cannot handle sarcasm data. So, in future this work can be enhanced to effectively handle sarcasm by involving additional rules. Measures should be taken to overcome the stale information in order to increase the prediction levels. To improve the decision-making process the root causes with low cost and time benefit should be given importance in the future.

References

1. Liu B (2012) Sentiment analysis and opinion mining. Synthesis lectures on human language technologies 5(1):1–167
2. Park D, Kim S (2009) The aspects of consumer knowledge on message processing of electronic word-of-mouth via online consumer reviews. Electron Commer Res Appl 7(4):399–410
3. Zhu F, Zhang X (2010) Impact of online consumer reviews on sales. The moderating role of product and consumer characteristics. J Mark 74(2):133–148
4. Hu Z, Hu J, Ding W, Zheng X (2015) Review sentiment analysis based on deep learning. In: Proceedings of the 12th IEEE international conference on e-Business engineering (ICEBE'15), pp. 87–94
5. Huang S, Liu X, Peng X, Niu Z (2012) Fine-grained product features extraction and categorization in reviews opinion mining. In: 12th international conference on data mining workshops (ICDMW), vol 6, pp 80–86
6. Rana TA, Cheah YN (2016) Aspect extraction in sentiment analysis: comparative analysis and survey. Artif Intell Rev 46(4):459–483
7. Hu M, Liu B (2004) Mining and summarizing customer reviews. In: Proceedings of the tenth ACM SIGKDD international conference on knowledge discovery and data mining. ACM, pp 168–177
8. Liu Q, Gao Z, Liu B, Zhang Y (2016) Automated rule selection for opinion target extraction. Knowl-Based Syst 104:74–88

9. Pang B, Lee L (2004) A sentimental education: sentiment analysis using subjectivity summarization based on minimum cuts. In: Proceedings of the 42nd annual meeting on association for computational linguistics. Association for Computational Linguistics, p 271
10. Pang B, Lee L (2008) Opinion mining and sentiment analysis. Found Trends® Inf Retr 2(1–2):1–135
11. Amplayo RK, Song M (2017) An adaptable fine-grained sentiment analysis for summarization of multiple short online reviews. Data Knowl Eng 110:54–67
12. Bagheri A, Saraee M, De Jong F (2013) An unsupervised aspect detection model for sentiment analysis of reviews. Nat Lang Process Inf Syst 140–151
13. Cambria E (2016) Affective computing and sentiment analysis. IEEE Intell Syst 31(2):102–107
14. Popescu AM, Etzioni O (2005) Extracting product features and opinions from reviews. In: Proceedings of EMNLP-2005, pp 3–28
15. Popescu AM, Nguyen B, Etzioni O (2005) OPINE: extracting product features and opinions from reviews. In: Proceedings of HLT/EMNLP on interactive demonstrations. Association for Computational Linguistics, pp 32–33
16. Li Y, Qin Z, Xu W, Guo J (2015) A holistic model of mining product aspects and associated sentiments from online reviews. Multimed Tools Appl 74(23):10177–10194
17. Eirinaki M, Pisal S, Singh J (2012) Feature-based opinion mining and ranking. J Comput Syst Sci 78(4):1175–1184
18. Marrese-Taylor E, Velásquez JD, Bravo-Marquez F (2014) A novel deterministic approach for aspect-based opinion mining in tourism products reviews. Expert Syst Appl 41(17):7764–7775
19. Lin W, Wu Z, Lin L, Wen A, Li J (2017) An ensemble random forest algorithm for insurance big data analysis. IEEE Access 5:16568–16575
20. Abualigah LM, Khader AT (2017) Unsupervised text feature selection technique based on hybrid particle swarm optimization algorithm with genetic operators for the text clustering. J Supercomput 73(11):4773–4795
21. Khan AZ, Atique M, Thakare VM (2015) Combining lexicon-based and learning-based methods for Twitter sentiment analysis. Int J Electron Commun Soft Comput Sci Eng (IJECSCSE) 89
22. Hai Z, Chang K, Cong G (2012) One seed to find them all: Mining opinion features via association. In: Proceedings of the 21st ACM international conference on information and knowledge management, pp 255–264
23. Ma B, Zhang D, Yan Z, Kim T (2013) An LDA and synonym lexicon based approach to product feature extraction from online consumer product reviews. J Electron Commer Res 14(4):304–314
24. Liu K, Xu L, Zhao J (2015) Co-extracting opinion targets and opinion words from online reviews based on the word alignment model. IEEE Trans Knowl Data Eng 27(3):636–650
25. Yan Z, Xing M, Zhang D, Ma B (2015) EXPRS: An extended pagerank method for product feature extraction from online consumer reviews. Inf Manag 52(7):850–858
26. Samha AK, Li Y, Zhang J (2014) Aspect-based opinion extraction from cus- tomer reviews. arXiv preprint arXiv:1404.1982
27. Wang T, Cai Y, Leung HF, Lau RY, Li Q, Min H (2014) Product aspect extraction supervised with online domain knowledge. Knowl-Based Syst 71:86–100
28. Wu C, Wu F, Wu S, Yuan Z, Huang Y (2018) A hybrid unsupervised method for aspect term and opinion target extraction. Knowl-Based Syst 148:66–73
29. Cruz FL, Troyano JA, Enríquez F, Ortega FJ, Vallejo CG (2013) Long autonomy or long delay. The importance of domain in opinion mining. Expert Syst Appl 40(8):3174–3184
30. Chen L, Qi L, Wang F (2012) Comparison of feature-level learning methods for mining online consumer reviews. Expert Syst Appl 39(10):9588–9601
31. Yang B, Cardie C (2013) Joint inference for fine-grained opinion extraction. In: Proceedings of the 51st annual meeting of the association for computational linguistics, pp 1640–1649
32. Li S, Wang R, Zhou G (2012) Opinion target extraction using a shallow semantic parsing framework. In: Proceedings of the twenty-sixth AAAI conference on artificial intelligence, pp 1671–1677

33. Li Y, Wang H, Qin Z, Xu W, Guo J (2014) Confidence estimation and reputation analysis in aspect extraction. In: 2014 22nd international conference on pattern recognition (ICPR). IEEE, pp 3612–3617
34. Poria S, Cambria E, Gelbukh A (2016) Aspect extraction for opinion mining with a deep convolutional neural network. Knowl-Based Syst 108:42–49
35. Das H, Naik B, Behera HS (2019) Medical disease analysis using neuro-fuzzy with feature extraction model for classification. Inform Med Unlocked 100288
36. Das H, Naik B, Behera HS (2020) An experimental analysis of machine learning classification algorithms on biomedical data. In: Proceedings of the 2nd international conference on communication, devices and computing. Springer, Singapore, pp 525–539
37. Das H, Naik B, Behera HS (2018) Classification of diabetes mellitus disease (DMD): a data mining (DM) approach. In: Progress in computing, analytics and networking. Springer, Singapore, pp 539–549
38. Jeyapriya A, Selvi CK (2015) Extracting aspects and mining opinions in product reviews using supervised learning algorithm. In: 2015 2nd international conference on electronics and communication systems (ICECS). IEEE, pp 548–552
39. He R, Lee WS, Ng HT, Dahlmeier D (2017) An unsupervised neural attention model for aspect extraction. In: Proceedings of the 55th annual meeting of the association for computational linguistics, long papers, vol 1, pp 388–397
40. Paredes-Valverde MA, Colomo-Palacios R, Salas-Zárate MDP, Valencia-García R (2017) Sentiment analysis in Spanish for improvement of products and services: a deep learning approach. Scientific Programming
41. Ruder S, Ghaffari P, Breslin JG (2016) INSIGHT-1 at SemEval-2016 Task 5: deep learning for multilingual aspect-based sentiment analysis. In: Proceedings of the 10th international workshop on semantic evaluation (SemEval ′16). San Diego, CA, USA, pp 330–336
42. Das H, Dey N, Balas VE (eds) (2019) Real-time data analytics for large scale sensor data. Academic Press
43. Dey N, Das H, Naik B, Behera HS (eds) (2019) Big data analytics for intelligent healthcare management. Academic Press
44. Dey N, Ashour AS, Kalia H, Goswami R, Das H (2019) Histopathological image analysis in medical decision making. IGI Global, Hershey, PA, pp 1–340. https://doi.org/10.4018/978-1-5225-6316-7
45. Pradhan C, Das H, Naik B, Dey N (2018) Handbook of research on information security in biomedical signal processing. IGI Global, Hershey, PA, pp 1–414
46. Sahoo AK, Mallik S, Pradhan C, Mishra BSP, Barik RK, Das H (2019) Intelligence-based health recommendation system using big data analytics. In: Big data analytics for intelligent healthcare management. Academic Press, pp 227–246
47. Sahoo AK, Pradhan C, Das H (2020) Performance evaluation of different machine learning methods and deep-learning based convolutional neural network for health decision making. In: Nature inspired computing for data science. Springer, Cham, pp 201–212
48. Zhang X, Chen F, Huang R (2018) A combination of RNN and CNN for attention-based relation classification. Procedia Comput Sci 131:911–917
49. Zheng X, Chen H, Xu T (2013) Deep learning for Chinese word segmentation and POS tagging. In: Proceedings of the 2013 conference on empirical methods in natural language processing, pp 647–657
50. Collobert R, Weston J (2008) A unified architecture for natural language processing: Deep neural networks with multitasks learning. In: Proceedings of the 25th international conference on machine learning. ACM, pp 160–167
51. Majumder N, Poria S, Gelbukh A, Cambria E (2017) Deep learning-based document modeling for personality detection from text. IEEE Intell Syst 32(2):74–79
52. Asghar MZ, Khan A, Zahra SR, Ahmad S, Kundi FM (2017) Aspect-based opinion mining framework using heuristic patterns. Clust Comput 1–19
53. Akhtar MS, Gupta D, Ekbal A, Bhattacharyya P (2017) Feature selection and ensemble construction: a two-step method for aspect-based sentiment analysis. Knowl-Based Syst 125:116–135

54. Das H, Jena AK, Nayak J, Naik B, Behera HS (2015) A novel PSO based back propagation learning-MLP (PSO-BP-MLP) for classification. In: Computational intelligence in data mining, vol 2. Springer, New Delhi, pp 461–471
55. Qiu G, Liu B, Bu J, Chen C (2011) Opinion word expansion and target extraction through double propagation. Comput Linguist 37(1):9–27
56. Hu M, Liu B (2004) Mining opinion features in customer reviews. AAAI 4(4):755–760
57. Scaffidi C, Bierhoff K, Chang E, Felker M, Ng H, Jin C (2007) Red Opal: product-feature scoring from reviews. In: Proceedings of the 8th ACM conference on electronic commerce. ACM, pp 182–191
58. Rana TA, Cheah YN (2017) A two-fold rule-based model for aspect extraction. Expert Syst Appl 89:273–285
59. Kang M, Ahn J, Lee K (2018) Opinion mining using ensemble text hidden Markov models for text classification. Expert Syst Appl 94:218–227
60. Kang Y, Zhou L (2017) RubE: rule-based methods for extracting product features from online consumer reviews. Inf Manag 54(2):166–176
61. Hai Z, Chang K, Kim JJ, Yang CC (2014) Identifying features in opinion mining via intrinsic and extrinsic domain relevance. IEEE Trans Knowl Data Eng 26(3):623–634
62. Wang W, Xu H, Wan W (2013) Implicit feature identification via hybrid association rule mining
63. Munir K, Anjum MS (2018) The use of ontologies for effective knowledge modelling and information retrieval. Appl Comput Inform 14(2):116–126
64. Rout M, Jena AK, Rout JK, Das H (2020) Teaching–learning optimization based cascaded low-complexity neural network model for exchange rates forecasting. In: Smart intelligent computing and applications. Springer, Singapore, pp 635–645
65. Sun Q, Niu J, Yao Z, Yan H (2019) Exploring eWOM in online customer reviews: Sentiment analysis at a fine-grained level. Eng Appl Artif Intell 81:68–78
66. Lau RY, Li C, Liao SS (2014) Social analytics: learning fuzzy product ontologies for aspect-oriented sentiment analysis. Decis Support Syst 65:80–94
67. Konys A (2018) Towards knowledge handling in ontology-based information extraction systems. Procedia Comput Sci 126:2208–2218
68. Geetha R, Rekha P, Karthika S, (2018) Twitter opinion mining and boosting using sentiment analysis. In: 2018 international conference on computer, communication, and signal processing (ICCCSP). IEEE, pp 1–4
69. Hazim M, Anuar NB, Ab Razak MF, Abdullah NA (2018) Detecting opinion spams through supervised boosting approach. PLoS ONE 13(6):e0198884
70. Wang G, Zheng D, Yang S, Ma J (2017) FCE-SVM: a new cluster based ensemble method for opinion mining from social media. Inf Syst e-Bus Manag 1–22
71. Jain AP, Katkar VD (2015) Sentiments analysis of Twitter data using data mining. In: 2015 international conference on information processing (ICIP). IEEE, pp 807–810
72. Hu K, Wu H, Qi K, Yu J, Yang S, Yu T, Zheng J, Liu B (2018) A domain keyword analysis approach extending Term Frequency-Keyword Active Index with Google Word2Vec model. Scientometrics 114(3):1031–1068

Chapter 6
Spider Monkey Optimization Algorithm in Data Science: A Quantifiable Objective Study

Hemant H. Kumar, Tanisha Sabherwal, Nimish Bongale, and Mydhili K. Nair

Abstract This chapter addresses an algorithm, namely Spider Monkey Optimization (SMO) for the researchers interested in optimization, swarm intelligence is an engaging realm. Several nature-inspired computing techniques and algorithms have been evolved by simulating the swarming traits of creatures such as ants, fishes, honey-bees, fireflies, etc. The results put forth are indeed extremely promising when applied to find the best-fit solution to a real-world optimization problem. Spider Monkey Optimization Algorithm (SMO) is detailed which is modeled on the basis of the social behavior of spider monkeys. This chapter focuses on detailing the nuances of SMO specifically the phases involved, namely the Leader Phase, Learning Phase, and Decision Phase. It also aims to give an introduction to the basic mathematical jargon and fundamentals that are required to model an SMO algorithm for finding the best-fit solution to an in-hand problem. Various variants of SMO are also covered in this chapter with a detailed overview of the pros and cons of each of the variants focusing on the research gaps. A wealth of information in the form of practical use-cases and implementation nuances is provided in this section, which will be a ready reckoner for a researcher reader aiming to flatten the learning curve associated with experimenting on basic SMO as well as trying out its variants for optimization problems in their specific realm of research. Further, the chapter gives an overview of a transformed version of SMO, Binary SMO with a real-world application and its comparison against other optimization techniques. It also comprises a case study of SMO where it is applied to Traveling Salesman Problem (TSP). SMO is used for this optimization problem to find the least travel cost covering all the cities. The chapter concludes by giving the reader an insight into the latest research trends in SMO.

Keywords SMO—Spider monkey optimization · Nature-inspired technique · Swarm intelligence · Variants of SMO

H. H. Kumar
School of Computer Science and Engineering, Vellore Institute of Technology, Vellore, Tamil Nadu 632014, India

T. Sabherwal · N. Bongale · M. K. Nair (✉)
Department of Information Science and Engineering, Ramaiah Institute of Technology, Bangalore 560054, India
e-mail: mydhili.nair@gmail.com

J. K. Rout et al. (eds.), *Machine Learning for Intelligent Decision Science*, Algorithms for Intelligent Systems, https://doi.org/10.1007/978-981-15-3689-2_6

1 Spider Monkey Inspired Technique

Spider Monkey Optimization (SMO), a nature-inspired technique to resolve optimization problems, with its core principle based on the social behavior of swarms, primarily Spider Monkeys (SMs), refer Fig. 1, is a very recently developed technique by Bansal et al. [1]. These creatures follow a fission–fusion societal behavior for their survival [1]. Spider Monkeys inherently have this "Fission–Fusion Social Structure (FFSS)" while scouting for the most suitable food source for them. In such a structure, a population of spider monkeys "fission" into smaller subgroups with one as a Global Leader. Each and every one of these subgroups, steered by a Local Leader, forage in different directions for adequate provision of food for all the members of the swarm reducing foraging competitivity. Refer Fig. 1 for pictorial representation.

The course of scrounging for food steers ahead through a communication flow within and outside these subgroups. The behavior pattern of all these spider monkeys depends on their instinctive communication link with one other. These faunal species, therefore, act according to the changes in the environment and the motion of other group members. After the required attainment of food, all the subgroups come together to unite. This process is usually referred to as "Fusion".

1.1 Introduction to Spider Monkey Optimization Algorithm (SMO)

Spider Monkey Optimization (SMO) is a fairly recent swarm intelligence algorithmic optimization technique. SMO was developed by Bansal et al. [1] by studying the astute societal behavior of spider monkeys when hunting for food.

The food-hunting manners of animals such as spider monkeys with FFSS characteristic are split into the following steps [1]:

(a) Defining the population size for a given set of spider monkeys and commence the hunting for food by evaluating the approximate farness from the food source location.

Fig. 1 Spider monkeys when foraging food

(b) The population is split up into groups and its members reposition themselves depending on the distance from the sources of food. These distances are re-evaluated iteratively.
(c) The Local Leader, hereby referred to as LL, assesses these distances sent by the group members and upgrades its foremost position within the group. In case this "foremost" location is not upgraded for a particular number of iterations, all the group members begin scouting for food in multiple directions.
(d) The Global Leader, hereby referred to as GL, evaluates the foremost position sent by all the LLs and updates its ever best position. Again, if this "best ever position" is not updated, it is assumed that the group is stagnating and the GL divides it into smaller sized groups, thus forming subgroups.

These steps are ceaselessly implemented until the desired goal is attained. In order to help local and GLs take suitable decisions, two control parameters are needed, namely "*GlobalLeaderLimit*" and another is "*LocalLeaderLimit*" to supply directions to global and local leaders. These are explained in detail in Sect. 2, while the algorithm is given in Sect. 3.

In the next section, a real-world case study is described to make the reader realize the importance of optimization techniques in practical scenarios. This case study is chosen as it applies Spider Monkey Optimization technique with good performance results.

1.2 Case Study: SMO to Optimize Golestan and Voshmgir Dam Operations in Iran

Dams and reservoirs can rightfully be called a mode of survival in areas prone to droughts and water scarcity. The prime need in such cases is optimizing the design of the dam in order to provide minimal overhead and production cost as well as maximal utilization of resources. Optimization algorithms are applied to solve the problem iteratively until a desired optimal output is achieved. SMO has proved to be ideally suited for nonlinear constrained optimization problems, hence in order to lessen the irrigation deficits, it is been suggested for the complicated modeling of this two-reservoir system in Iran [2].

This design of the dam with maximum cost-effectiveness demands the inclusion of computational methods to exclude negative aspects with exploration–exploitation trade-off [2]. The idea of "exploration–exploitation" essentially involves finding all the possible solutions to a problem from a search space followed by refining the same to achieve the optimal result. In such a scenario, where the optimal solution is to be searched for from a given set of all possible values, SMO becomes the best bet for optimization of Golestan and Voshmgir dams as it includes both the exploration and exploitation aspect. An experimental analysis of SMO against other optimization techniques stated as follows [2]:

Table 1 Calculated results for algorithms applied to Golestan dam

Algorithm	Million cubic meters (MCM) of water
PSA—Particle Swarm Algorithm	6.7
Standards operation policy method	19
GA—Genetic Algorithm	14.6
Water Cycle Algorithm	4.1
Harmony Search Algorithm	16.4

- Maximum deficits for the Golestan dam as computed by spider monkey algorithm 2.1 and 1.9 million cubic meters (MCM) of water, i.e., the utilization of water for irrigation is minimized by applying SMO. The value calculated for other algorithms is given in the table below (Table 1).
- The overall time taken by the spider monkey algorithm for computation is 50 s while the Genetic and Particle Swarm Algorithm (PSA) takes 172.6 s and 117.4 s, respectively, to optimize the problem.
- Loss functions such as Mean Absolute Error (MAE), Root mean square error (RMSE) between the required and actual million cubic meters (MCM) of water released by the Golestan and Voshmgir dams, have the minimum values for SMO among the other evolutionary algorithms deployed and their performance evaluated.
- The volumetric reliability index was calculated as 95 and 97% by SMO for the dams which are the criterion for water supply resources. Thus, it confirms the high computational effectiveness of this algorithm.

The above result obtained from the experimental analysis by Ehteram proves that SMO, in comparison to other primitive algorithms, has an exploration–exploitation feature which is extremely helpful for the appropriate optimization of dams and reservoir systems.

2 Introduction to Mathematical Modeling and Implementation of Spider Monkey Optimization Algorithm (SMO)

There are several phases in SMO, namely Local Leader Phase (LLP), Global Leader Phase (GLP), and Local Leader Learning Phase (LLLP).

2.1 Population Initialization

To start with, the SMO produces a consistent yet disseminated population of spider monkeys (SM's). Let us assume we have N spider monkeys, where every individual is portrayed as a D-dimensional vector. Every dimension *j* of the SM is initialized in the following manner, as proposed by Agrawal et al. [19]:

$$SM_{ij} = SM_{minj} + rand[0, 1](SM_{maxj} - SM_{minj}) \tag{1}$$

where maxj and minj are the upper and lower limits in the j direction while rand[0,1] is a random number generator function lying in the given range (Fig. 2).

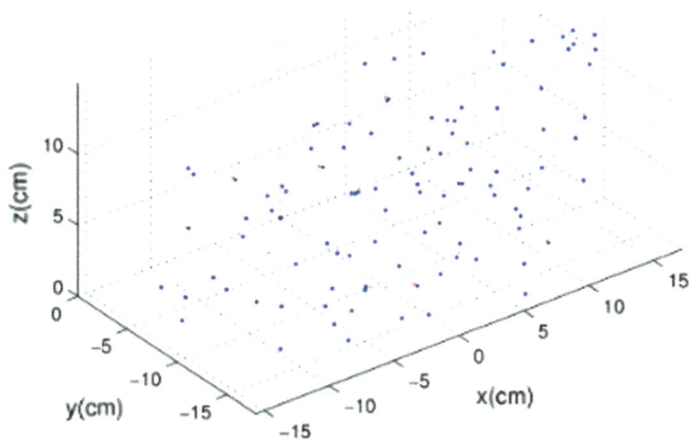

Legend

Blue dots:- SM positions in 3D vector space, measured wrt origin

X axis:- General x axis with (-15, 15) interval, representing position along x direction

Y axis:- General y axis with (-15, 0) interval, representing position along y direction

Z axis:- General z axis with (0, 10) interval, representing position along z direction

Fig. 2 SM population initialization

2.2 Local Leader Phase

In the Local Leader Phase, acquaintances and experiences of the LL and the members are taken into account and every spider monkey upgrades its location based on certain parameters. A new fitness value is thus obtained. If this new location's fitness value is of more importance than before, then the spider monkey is said to "swing about" and change its position. The update equation is given by

$$SM_{newij} = SM_{ij} + rand[0, 1](LL_{kj} - SM_{ij}) + rand[-1, 1](SM_{rj} - SM_{ij}) \quad (2)$$

where the location of the LL is given by LL_{kj}, and SM_{rj} represents the dimension "j" individual "r", an arbitrary selection made from the group (Fig. 3).

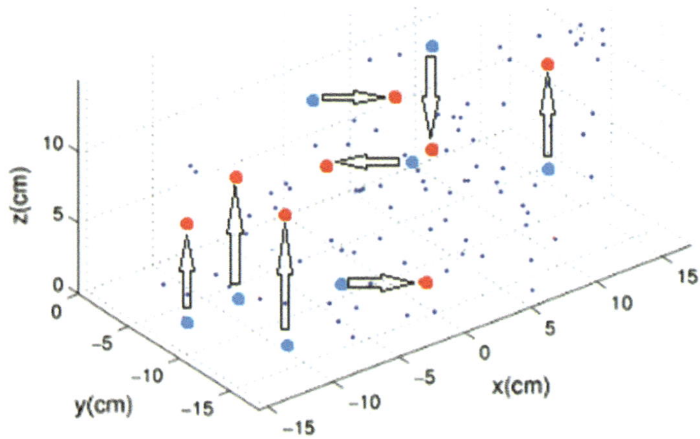

Legend

Blue dots with smaller radii:- SM unchanged positions in 3D vector space, measured wrt origin

Blue dots with larger radii:- Positions of SM to be changed

Red dots:- Updated SM positions

X axis:- General x axis with (-15, 15) interval, representing position along x direction

Y axis:- General y axis with (-15, 0) interval, representing position along y direction

Z axis:- General z axis with (0, 10) interval, representing position along z direction

Fig. 3 Population reorganization during (LLP) Local Leader Phase

2.3 Global Leader Phase

The Global Leader Phase (GLP) comes right after Local Leader Phase (LLP) described above. The SM's check their fitness by exploiting the previous experience of GL and other monkeys of the local groups. The location upgrade equation of this particular stage differs slightly

$$SM_{newij} = SM_{ij} + rand[0, 1](GL_j - SM_{ij}) + rand[-1, 1](SM_{rj} - SM_{ij}) \quad (3)$$

where GL_j is the location of the GL while j {1, 2, …, D} is an index that is randomly chosen. The locations of spider monkeys (SM_i) are re-evaluated depending on probability conditions. Fitness plays a major role in deciding the probability. In this technique, a fitter individual will have better chances of making it to the next level. An equation giving this probability can be written as follows:

$$prob = 0.9 \times \frac{fitness}{max\ fitness} + 0.1 \quad (4)$$

2.4 Global Leader Learning Phase

The spider monkey having the fittest position is chosen and in comparison to the GL. If the SM is found to have characteristics comparable to the GL (whose position may not be updating), then Global Leader Count (GLC) is stepped up by one.

2.5 Local Leader Learning Phase

Using the greedy technique, the LLs position is updated. This position is contrasted with the old one. If there is no noticeable change, and there is an SM in the local group as fit as the LL, then Local Leader Count (LLC) is stepped up by one.

2.6 Local Leader Decision (LLD) Phase

If there is a deadlock in terms of changes in the Local Leader Positions (LL), then a decision must be taken as to how to ascertain the positions of the SM's in the individual groups. This can be performed at random or by the knowledge of GL and LL. An equation can thus be written on this basis [19]:

$$SM_{newij} = SM_{ij} + rand[0, 1](GL_j - SM_{ij}) + rand[0, 1](SM_{ij} - LL_{kj}) \quad (5)$$

2.7 Global Leader Decision (GLD) Phase

Problem arises when the position of the global leader is not updating as per requirements and is not meeting the Global Leader Limit mark. A decision thus needs to be taken, as a consequence of which the group either splits into small batches or merges into a large batch.

If the group splits into small batches, then a new Local Leader Learning process needs to be initiated in order to select LLs at the sub-group level.

Fusion occurs if the max number of batches is already reached, while the position of the GL is not updated, then fusion occurs, thus merging all the smaller batches into a large one.

Perturbation rate (p_r), as shall be seen in the following algorithms, is the degree of movement or randomness of a certain entity (here, SMO) with respect to its surroundings.

Probability (prob_i) is used in the Global Leader Phase to determine which individual makes it to the next level, with a fitter individual given better chances to do so (Refer Eq. 4 in Sect. 1.2).

3 Implementation of the Mathematical Model via Algorithms (Local Leader)

Algorithm for Local Leader Phase Position Updation(LLPU)

```
1. function LLPU
2. for every lth group (till Maximum Group Size) do
3.     for every Spider Monkey belonging to lth group do
4.        for randomly chosen index belonging to {1,2,...D} do
5.           check condition if R(0, 1) greater than Perturbation Rate
(pr) then
6.           Apply Eq(2)to calculate SMnew_xy
```

```
Algorithm for Local Leader Decision(LLD)Phase
function LLD
for every lth group (till Maximum Group Size) do
   if local-limit-countk > local-leader-limit then
   Initialize local-limit-countk to zero
      Assign GS(GroupSize) to the lth GS
     for x ← 1 to Group Size do
        for each y in { 1, 2,...D }do
          if R(0, 1) is greater than perturbation rate 'pr' then
          SMnewxy ← SMminy + R(0, 1) × (SMmaxy -SMminy)
          else
           Apply Eq(5)to calculate SMnewxy
```

4 Implementation of Mathematical Model via Algorithms (Global Leader)

```
Algorithm for Global Leader Position Updation(GLPU)
function GLPU
for every lth group (till Maximum Group Size) do
   Initialize counter to 1
   Initialize GroupSize (GS) to lth GroupSize
   while counter is less than GroupSize do
   for x ← 1 till GroupSize (GS) do
      if R(0, 1) is less than probi then
        Increment counter (counter++)
        Randomly choose y from set {1,2,...D}
        Randomly choose SMr from lth group so that value of r & i are equal
   Apply Eq(3)to calculate SMnewxy
```

```
Algorithm for Global Leader Decision (GLD) Phase
1. function GLDP
2. if glc(globallimitcounter) is greater than
gllc(globalleaderlimitcount)then
3.    Initialize glc to zero
4.    if No. of Groups is lesser than Max Group Size (MG) then
5.          Divide Groups
6.    else
7.           Fuse all groups into one single group
8. Update Local Leader positions
```

5 Variants of Spider Monkey Applied to Data Science

5.1 Constrained Spider Monkey Optimization (SMO) Algorithm

CSMO, a variant of the SMO as stated Hazrati et al. "Modified spider monkey optimization" [21], optimizes problems by using Deb's technique (discussed in the following section in detail) for handling constraints. In this section, the performance of CSMO is investigated and the results are compared and contrasted with similar constrained versions of older optimization algorithms such as Genetic Algorithms (GA), Artificial Bee Colony (ABC) and Particle Swarm Optimization (PSO) [25]. The experimental results prove that CSMO is a great candidate when it comes to the task of global optimization for constrained optimization problems. Thus, the CSMO sets a benchmark for other optimization methodologies when it comes to solving real-world problems.

5.1.1 Handling Constraints Using Deb's Technique

Deb's technique is meta-heuristic method by which constraints can be applied to a problem. It has the following three rules under it:

- The best solution is picked based on feasibility.
- If there are two equally viable solutions, the one with greater fitness is chosen.
- If there are two infeasible solutions (and none note worthily feasible), the one having less violation of constraints is picked.

An equation fits these conditions in our case:

$$\mathrm{viol_i} = \sum_{i=1}^{p} G_\mathrm{i}(X) + \sum_{j=p+1}^{m} H_\mathrm{j}(X) \tag{6}$$

Initialization of Population

In CSMO, the uniform distribution is put to use to generate the SM population with certain lower and upper bounds. This thus accepts feasible and infeasible solutions as well. If this is not done, and only the feasible ones are selected, then there might be an overshoot in terms of computational power consumption.

Local Leader Phase—LLP in CSMO

Local Leader (LLP) Phase here is similar to in SMO (as mentioned earlier) except for the choosing of result for contrasting. In CSMO, Deb's rules are the underlying technique used to upgrade the position of spider monkeys. The updated location and present location of an SM are contrasted in accordance to these rules, and then the necessary steps are performed.

Global Leader Phase—GLP in CSMO

In GLP of SMO, a given solution for GL decision is upgraded on the basis of its "prob_i" value (mentioned in the algorithm in Sect. 1.4). In CSMO, the fitness of every result is sought such that a viable solution always gets a higher priority over the infeasible one.

Global Leader Learning Phase—GLLP in CSMO

A result having the maximum fitness value is found. Then, it is contrasted with the GL of the swarm using Deb's rules. If the present GL is inferior to the best member of the swarm then the best monkey is assigned the GL, and a position update occurs; otherwise, GL retains its position.

Local Leader Learning Phase—LLLP in CSMO

This phase is similar to that in the basic SMO, apart from the fact that the comparison is performed based on Deb's rule rather than the straightforward fitness.

Local Leader Decision Phase—LLDP in CSMO

In this step, if the LL has not changed position, the spider monkey group is re-initialized. Although this step may cause more infeasible solutions to occur, it is a necessary one. The capacity of the algorithm to scrounge remains unhindered.

Global Leader Decision Phase—GLDP in CSMO

This step is exactly the same as in the basic version of SMO, where if the GL hasn't traversed any distance, a decision must be taken in order to undergo fission or fusion.

5.2 Ageist Spider Monkey Optimization

The so-called "intelligent" SM behavior is primarily due to the fact that their foraging process is done by fission and fusion. The SM's showcase the skills of an intelligent life form. However, we don't take into account the factors affecting the search capacity of every monkey, for example age.

The ageist SMO is designed based on this very principle. The younger spider monkeys will be swift and more speedy when it comes to interacting and updating their locations, while on the other hand, the older ones will be slower. These faster monkeys will interconnect and upgrade their locations before the slow-moving monkey and will endow them with better leaning. The ASMO thus works on age and dynamic differences between SM's. Thus, further groups need to be formed in order to separate the group by biological similarities. All the SM's in this small group will have the same or similar level of abilities. The best mini-group updates its position first, giving the best possible learning experience to the other groups, and we thus move toward a better optimal solution.

5.2.1 ASMO Algorithm

The concept, as discussed earlier, is to split a group of SM's into *M* number of smaller batches or subgroups, and this variable "*M*" can be assigned a value as per the will of the user. Instead of upgrading the locations of all individuals of the group and thence choosing a profitable location between the old and the present by using a greedy technique based on the fitness, the given steps are carried out for one individual small sub-group and then it jumps to the next smaller sub-group. ASMO implements an ageist strategy in only Local Leader Phase (LLP), as proposed by Sharma et al. [18]. Rest of the steps remain the same.

```
Algorithm for Updating the Position in ASMO (ASMOLLPU)
1. function ASMO_LLP
2. for every  l in set {1,2...MG} do
3.     for every l^th group (till Maximum Group Size) do
4.           for every member SMx in the mth mini-segmented-group do
5.                for every y in the set {1,2,...D} do
6.                      Check whether R(0, 1) greater than perturbation
rate pr then
7.                      Apply Eq(2)to calculate SMnew_xy
8. for every  member SMx ∈ mth mini-segmented-group do
9.     compute the fitness every single time
10.    if the new 'fitness'  is more than 'fitness' previously got then
11.          for every y in set {1,2,...D}do
12.          SMxy ← SMnewxy
13.    current fitness← fitness newx
```

5.3 Self-adaptive Spider Monkey Optimization (SaSMO)

Self-Adaptive Spider Monkey Optimization (SaSMO) is a cognition-based optimization algorithm. It is an efficient technique for complex optimization problems as compared to basic Spider Monkey Optimization. SaSMO proposes that the position updation is based on the former position of the agents. In every iteration, the current location of every spider monkey dwindles linearly from 100 to 50%. In SaSMO, the probability of former positions of every SM contributes to its new position in the group.

The underlying idea behind the Self-Adaptive SMO is the presumption that the LL-phase under successive iterations might near the optimal solution to be obtained. The above assumption is considering that in the first iteration, the output will be distant from the desired solution. Hence, this optimization is based on the probability fitness of every spider monkey during the generation of a new position for the monkey.

5.3.1 Stages of the SaSMO

The initialization in the SaSMO is essentially similar to the basic SMO. The initial locations of all the spider monkeys are computed using the same formula given in Sect. 1.2 and the greedy approach is applied in the subsequent steps.

Local Leader Phase

In this stage, where the locations of the monkeys (SMs) are computed depending on the experience of the LLs of the group, the probability of every individual is entailed. SaSMO's LL selection phase follows a greedy approach selection between the former and recent positions. Thus, this solution may not be the global optimum one. The new position is calculated using the following formula:

$$\mathrm{SMnew}_{m,n} = \mathrm{SM}_{m,m} + U(0,1) \times (\mathrm{LL}_{k,n} - \mathrm{SM}_{m,n}) + p_m \times (\mathrm{SM}_{p,n} - \mathrm{SM}_{m,n}) \tag{7}$$

where p_m denotes the mth solution which is the probability of every spider monkey and is computed by employing the fitness value of the individual SM [3].

Global Leader Phase

In this phase, the computation of the locations of the spider monkeys (SM's) depends on the prior experience of the LL's of the group. The probability of each monkey is calculated using the formula:

$$p_m = 0.9 \times \frac{\text{fitness}_m}{\text{fitness}_m ax} + 0.1 \tag{8}$$

The same greedy selection is applied. The new position is calculated using a similar formula [3]:

$$\text{SMnew}_{m,n} = \text{SM}_{m,n} + U(0,1) \times (\text{GL}_n - \text{SM}_{m,n}) + p_m \times (\text{SM}_{p,n} - \text{SM}_{m,n}) \tag{9}$$

Position Update of Local Leader—(LL)

The computation of the location of LL-Local Leader is on the basis of a constant ω using the above-stated assumption that the location of the LL linearly decreases from 100 to 50%. Hence, the value for ω_{max} and ω_{min} is set to 1 and 0.5, respectively. The updation of the location of the LL is based on the following steps [3].

If the updation of the location of LL does not occur after the Local Leader Limit (LL_{limit}) then,

If rand[0, 1] > perturbation rate(p_r)

$$\text{SMnew}_{m,n} = \text{SM}_{minn} + U(0,1) \times (\text{SM}_{\text{maxj}} - \text{SM}_{\text{minn}}) \tag{10}$$

else

$$\text{SMnew}_{m,n} = \text{SM}_{m,n} + U(0,1) \times (\text{GL}_j - \text{SM}_{m,n}) + (\text{SM}_{m,n} - \text{SM}_{k,n}) \times \left(\omega\text{max} - \frac{\text{iter}}{\text{max} - \text{iter}}(\omega\text{max} - \omega_{\text{min}})\right) \tag{11}$$

The position computation is similar to the basic SMO given in Sect. 1.2.

5.3.2 Algorithm for SaSMO

Initialization stage

Define population size with LLL-Local Leader Limit & GLL-Global Leader Limit

Define the largest no. of groups.

State 'Pr' : Perturbation Rate.

Load Binary Optimization Objective function.

Produce a value for every spider monkey randomly using the generic equation (refer section 1.2).

Compute the fitness value.

Designate the LL and GL by greedy search.

while termination case is not attained do

LOCAL LEADER PHASE

Produce new values using the Eq. (7).

Compute fitness for the new values obtained

Determine the better solution based on the fitness obtained

GLOBAL LEADER PHASE

Compute probabilities using Eq. (8)

Produce new values based on Eq. (9)

Compute fitness of new solution

Determine the better solution based on the fitness obtained

Change the positions of the Global(GL) and LL

LOCAL LEADER DECISION PHASE

If the updation of position of the LL doesn't occur after the LL Limit then do

    if rand[0,1]>pr

        Compute the locations using Eq. (10)

    else

        Use Eq. (11)

GLOBAL LEADER DECISION PHASE

If the updation of the location of GL doesn't occur even after GLL_{limit} then do

    Check wether $Global - Limit - Count > GLL_{limit}$) then

        Set $Global - Limit - Count$ = 0

    Check whether (no. of groups $<$ MG-maximum no of groups)) then

        Split the community into batches/subgroups

    else

        Unify all the batch of monkeys(SMs)

Update the Local Leader Position

end while

Hence, the above algorithm is a refined version of basic SMO, wherewith every iteration, the solution is expected to near the optimal solution. It can be summarized in the form of the algorithm given above. This states that in the Local Leader Phase, after the first iteration, every successive iteration aims to achieve a more optimal solution than the previous one. Unlike the basic SMO where every step taken by the spider monkeys is uniform, SaSMO allows the individuals to take nonuniform strides in order to arrive at the best solution in lesser time. With time or with every iteration, each spider monkey takes a smaller leap toward the food source but the distance between the food source and the monkey is diminished. In this case, the leap after the first iteration is the maximum.

The basic approach for this algorithm is that the initial positions of the monkeys are very random and undesirable. These positions in the later stages turn out to be more reliable with an increase in the total iterations. With decreased strides, the position of the spider monkeys meander in the proximity of the best solution and these solutions can be called "better" solutions since they near the optimal result required.

5.4 *Chaotic Spider Monkey Optimization*

The basic Spider Monkey Optimization (SMO) algorithm consists of a set of control parameters which paves the way for the optimization process through every iteration. Perturbation rate is a prime control parameter that affects the algorithm in a number of ways. It is, in most cases, a linearly increasing function. For such a perturbation rate, the performance of the Spider Monkey Optimization might not be applicable to domains involving a nonlinear function. Hence, a revised version of SMO called the Chaotic Spider Monkey Optimization (CSMO) is proposed wherein the perturbation rate has been modified for better performance and global optima search in nonlinear objective functions.

SMO falls in the category of meta-heuristic algorithm, hence, exploration and exploitation [1] is one of the two principal phases involved. With every iteration, SMO explores the position of each spider monkey and exploiting the result of the same, with a linear increase in the perturbation rate. However, experimental analysis suggests that as the number of iteration increases in SMO, the result is drawn more toward the local optima due to low rate of convergence. To solve this problem, in the following algorithm the perturbation rate varies "chaotically", i.e., nonlinear and divergent. Such an algorithm is expected to show better results for global optimization.

For the proposed Chaotic Spider Monkey Optimization, as mentioned by Kumar et al. in "Automated soil prediction using bag-of-features and chaotic spider monkey optimization algorithm" [20], a chaotic function is formulated for every successive iteration. This function forms the basis of the perturbation rate which is to be increased nonlinearly. The function is given as follows [1]:

$$z_{(n+1)} = \mu \times z_n \times (1 - z_n)$$

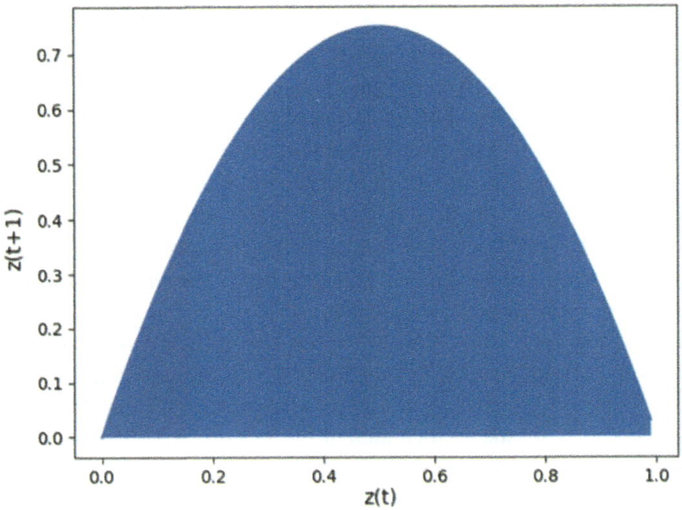

Fig. 4 Graph for the above-given function for perturbation rate

where z_n is the chaotic number of the nth which lies in the range [0,1] and μ is a constant. Figure 4 shows the nature of the following function for $\mu = 3$.

Hence the above Fig. 4 of the graph evidently shows that the perturbation rate depends on a variable called a "chaotic number" such that the change in its value is not linear. This change being nonlinear (parabolic in this case) makes it very suitable for optimizing nonlinear functions.

Using the above equation, the perturbation rate is varied using the formula [4]:

$$p_{(n+1)} = 1 - p_n \times \frac{\max \ - n}{\max} \times z \tag{12}$$

where p_n denotes the perturbation rate for nth iteration, max is the total iterations and n is the ongoing iteration. The value of p lies in the range [0,1].

The remaining steps in the generic SMO (as given in Sect. 1.2) remains the same for the Chaotic spider monkey Algorithm as well.

5.5 Hybridization of Genetic Algorithm and Spider Monkey Optimization

5.5.1 Genetic Algorithm

Genetic Algorithm (GA), developed by John Holland in 1960, is one of the most primitive approaches for solving optimization problems based on the evolution of the population. Along the lines of the naming of the following algorithm, a chromosome is a basic building block of the algorithm. Every chromosome inherits a set of genes, which in turn represent every variable of the problem. Finally, chromosome constitutes a set of real numbers that gives the solution to the optimization problem.

The process starts with randomly initializing the population with a finite number of individuals, every having its own value of fitness. Further, on the basis of the fitness value, new population arises using a set of genetic operations, namely, reproduction, crossover, and mutation.

- **Reproduction or Selection**: In this operation, parents regenerate offsprings, i.e., chromosomes selected from the population produce more individuals. Darwin's theory of evolution states that an individual with the maximum fitness value survives, hence, it is also called Selection operation. Therefore, in this process, a set of fitter population breeds a new set of individuals.
- **Crossover**: After the completion of the selection phase, crossover operator is applied. In this process, two individuals are chosen to breed better offspring. In this process, two individuals are randomly picked from a set of mating population and these exchange some genetic variables to procreate a new variable or individual with new traits. Some crossover used in Genetic Algorithms are Flat crossover [6], Arithmetic crossover, [7] and BLX-α crossover [8].
- **Mutation**: In this process, diversification of the population takes place. Succeeding the crossover, mutation is applied on the basis of a mutation probability. Usually, this mutation probability is set to a lower value.

5.5.2 Hybridization

Each of the two optimization algorithms belongs to the meta-heuristic category where exploration and exploitation are two primary aspects of the optimization technique process. The proposed algorithm can be divided into two phases, where every phase is governed by one of these two meta-heuristic algorithms. The first phase involves random initialization of the population and followed by the optimization of the problem using one of the above-stated algorithm (GA or SMO). In the other half, the output of the first acts as the input for this phase wherein it is optimized using the second algorithm (SMO or GA) [5].

Sequential Application of Spider Monkey Optimization and Genetic Algorithm (SMOGA)

SMOGA combines two algorithms where Spider Monkey Optimization is applicable in the first phase followed by Genetic Algorithm. A balance between the two optimization techniques is maintained for optimal results. Initially, the entire population is computed randomly followed by the updation of SMO until there is no progress for more than a hundred iterations. It is then followed by GA.

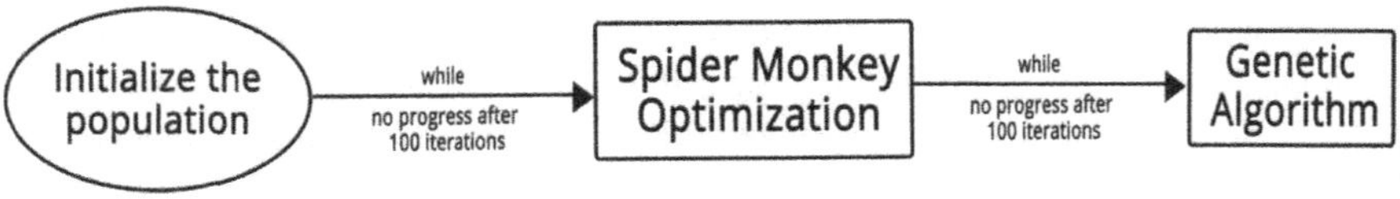


```
Run SMO
while (no progression after 100 iterations)
Run GA
while (no progression after 100 iterations)
```

Fig. 5 Sequential application of SMO-spider monkey optimization and GA-genetic algorithm

The first phase comprises the following six stages of SMO:

+ (LL) Local Leader Stage
+ (GL) Global Leader Stage
+ (GLL) Global Leader Learning Stage
+ (LLL) Local Leader Learning Stage
+ (LLD) Local Leader Decision Stage
+ (GLD) Global Leader Decision Stage

The control parameters in this stage are LocalLeaderLimit (LL_{limit}), GlobalLeaderLimit (GL_{limit}), Maximum Groups (MG), Perturbation Rate (p_r) as mentioned by Kavita Gupta and Kusum Deep—"Investigation of Suitable Perturbation Rate Scheme for Spider Monkey Optimization Algorithm" [22].

The next phase is further divided into five stages of GA, namely, Gene-pool Initialization, Gene-Selection, Gene-Crossover, Gene-Mutation, and Termination.

The initialization steps involve the initialization of the most effective output of the first stage which is fed as the input for the next stage. This is followed by gene-selection, gene-crossover, and gene-mutation. The following process reaches the termination stage when the finest optimal solution is got as the output for the second phase (Fig. 5).

Sequential Application of Genetic Algorithm and Spider Monkey Optimization (GASMO)

GASMO combines two algorithms where Genetic Algorithm is applicable in the first phase followed by Spider Monkey Optimization. A balance between the two optimization techniques is maintained for optimal results. Initially, the entire population is computed randomly followed by the updation of GA until there is no progress for more than a hundred iterations. It is then followed by SMO (Fig. 6).

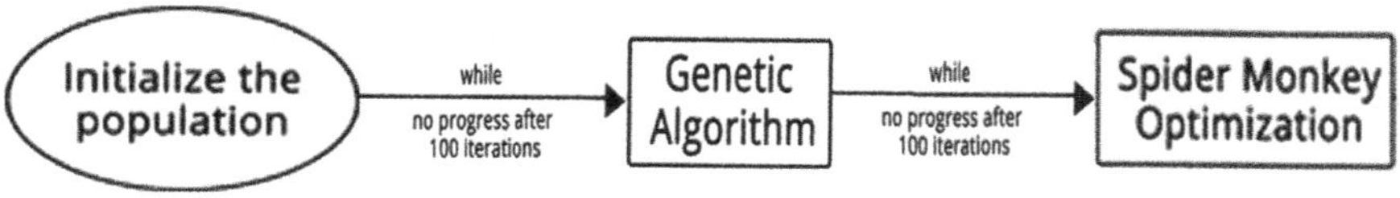


```
Run GA
while(no progression after 100 iterations)
Run SMO
while(no progression after 100 iterations)
```

Fig. 6 Sequential application of genetic algorithm and spider monkey optimization

6 Thinning of Circular Concentric Antenna Arrays (CCAA)

6.1 Antenna Array Optimization

For any electronic system, its heart lies at the capability of an antenna to receive and transmit electromagnetic signals between two entities that are not physically connected. The same when connected together, constitute an Antenna Array which functions as a single unit for the reception and transmission of radio waves. The radiation pattern primarily consists of two components, namely, the main beam and a set of smaller sidelobes. This pattern depends on the geometry of the structure and the arrangement of individual components in the space. For maximum and efficient utilization, the structure of these sidelobes is to be restrained and Antenna Arrays help in achieving the same. The individual antennae when arranged together, enhance the power radiated in a particular direction by superposing the electromagnetic waves transmitted.

The design of an Antenna Array is a strenuous problem that cannot be solved using primitive optimization techniques functioning on the basis of an initial guess. Hence, nature-inspired optimization techniques seem an exemplary approach. Genetic Algorithm [9, 10], Teaching Learning Based Optimization (TLBO) [13], Particle Swarm Optimization [11, 12], etc. have been applied for the design and implementation of Antennae Arrays for efficient wireless connection. The following section consists of comparison of Spider Monkey Optimization with all the above-stated.

Circular Concentric Antenna Arrays (CCAA) has its application in a number of fields including sonar, military, mobile, etc. Since its shape is circular, these Antenna Arrays are preferable due to easy transmission of waves in any desired direction and easier pattern synthesis of signals. Every side lobe associated with the radiation pattern of CCAA has a value called Side Lobe Level (SLL). This SLL is to be maintained at a minimal value since a higher value may result in unwanted interference, disrupting the transmission of signals. Thinning of CCAA is one of the methods of reducing SLL. It includes the exclusion of a set of antennae without affecting the radiation pattern of a completely occupied array. Hence, binSMO fits in perfectly where every active antenna is represented by "1" and inactive by "0".

6.2 *Binary Spider Monkey Optimization (BinSMO)*

With its roots emerging from Spider Monkey Optimization, binSMO solves binary optimization problems. Yuana [14] proposed an improved Particle Swarm Optimization designed for binary approach for optimization techniques and the following binSMO is inspired by the same. The binSMO proposes some changes in the generic SMO for its application in the binary sector. In order to initialize, a binary solution is applied making use of Bernoulli process. This is depicted in the equation shown in Eq. 13 below:

$$\mathrm{SM}_{m,n} = \begin{cases} 0, & \text{if}\, x < p \\ 1, & \text{otherwise} \end{cases} \tag{13}$$

where $\mathrm{SM}_{m,n}$ is the mth spider monkey located in the nth dimension, p is the assumed probability(considering 0.5 in this case) and x gets an arbitrary value in the range [0,1]. Hence, the above function states that if the random variable is less than the given probability (0.5), the dimension of the given spider monkey is zero and one, otherwise. Hence the following modifications in the SMO fit in perfectly for approaches involving binary optimization. After the following initialization, all the phases of a basic SMO are modified using logical operators (AND, OR and XOR) as follows.

6.2.1 Local Leader Phase

The spider monkeys in the group upgrades their respective positions based on the change in position of the members of the particular group and also the LL. The new position of the monkeys are using the following formula:

$$\mathrm{SMnew}_{m,n} = \begin{cases} \mathrm{SM}_{m,n} \oplus ((p \times ((\mathrm{LL}_{k,n} \oplus \mathrm{SM}_{m,n})) + (q \times (\mathrm{SM}_{r,n} \oplus \mathrm{SM}_{i,j}))), & \text{if rand} \geq p_r \\ \mathrm{SM}_{m,n}, & \text{otherwise} \end{cases} \tag{14}$$

On calculation of the fitness, better solutions are selected based on the comparison between the previous and present values.

6.2.2 Global Leader Phase

The spider monkeys in the group upgrades their respective position based on the change in position of the members of the particular group and also the GL. The updates position is based on an equation given as:

$$P_m = 0.9 * \frac{\text{fit}_m}{\max - \text{fit}} + 0.1 \tag{15}$$

where P_m is the probability, fit_m is the fitness value of the mth spider monkey and the max − fit is the greatest fitness value in the group of spider monkeys. Hence, using the position of the GL of the species, every of the position of the monkeys are updated using the following equation:

$$\text{SMnew}_{m,n} = \text{SM}_{m,n} \oplus ((p \times (\text{GL}_n \oplus M_{m,n})) + (q \times (\text{SM}_{r,n} \oplus \text{SM}_{i,j}))) \tag{16}$$

6.2.3 Local Leader Decision Phase

$$\text{SMnew}_{m,n} = \begin{cases} \text{SM}_{m,n} \oplus((p \times ((\text{LL}_{k,n} \oplus \text{SM}_{m,n})) + (q \times (\text{GL}_n \oplus \text{SM}_{i,j}))), & \text{if rand} \geq p_r \\ \text{SM}_{m,n}, & \text{otherwise} \end{cases} \tag{17}$$

In the following equation given above, $\text{SMnew}_{m,n}$ is the upgraded position of mth spider monkey in the nth dimension, $\text{SM}_{m,n}$ is the original position of mth spider monkey in the nth dimension, p and q are the arbitrary chosen variables in the range of [0,1], GL_n is the GL of the spider monkeys in the nth dimension and $\text{LL}_{k,n}$ is the LL of the kth group in the nth dimension. Also, $\oplus$, $\times$ and $+$ represent XOR, AND and OR, respectively.

6.2.4 Algorithm for BinSMO

The following is the algorithm for binSMO derived from the above equations:

```
Initialization stage

Define population size with values for LL Limit and GL Limit

Define the largest no. of batches.

State 'pr' : Perturbation rate.

Load Binary Optimization Objective function.

Produce a value for every spider monkey randomly using the binary values as
given is Eq. (13)

Compute the fitness value.

Designate the GL and LL  by greedy search.

while termination condition isn't reached do

LOCAL LEADER PHASE

Produce new values using the equation Eq. (14)

Compute fitness for the new values obtained.

Determine the better solution based on the fitness obtained

GLOBAL LEADER PHASE

Compute probabilities using equation Eq. (15)

Produce new values using the equation Eq. (16)

Compute 'fitness' value of the new result.

Determine the better result based on the value of 'fitness' obtained

Modify the positions of LL and GL.

LOCAL LEADER DECISION PHASE

if LL is stagnant

upgrade the location of all members on the basis of the equation Eq. (17)

end if

GLOBAL LEADER DECISION PHASE

if GL remains stagnant then split the group into two

while the largest number of groups are formed.

Merge all the groups into one.

Update LL.

end while

end if

end while
```

6.3 Geometry and Fitness Function of CCAA

6.3.1 Geometry of CCAA

Consider the following structure of CCAA, as given in Fig. 5. All the elements are positioned in M circular rings, every with the radius r_m. every of these M rings contain N elements denoted as N_i where i = 1, 2, 3, ..., n. Hence, the field pattern of this given CCAA can be denoted by the following formula [7, 8]:

$$E(\theta,\phi) = \sum_{i=1}^{M}\sum_{j=1}^{N_i} X_{mn} \times \exp\{b[kr_i \sin(\theta)\cos(\phi - \phi_{ij}) + \gamma_{mn}]\} \quad (18)$$

$$\gamma_{mn} = -kr_i \sin(\theta_0)\cos(\phi_0 - \phi_{ij}) \quad (19)$$

where k is the wavenumber given by $2\pi/\lambda$, λ is the wavelength of the radiation pattern. r_i denotes the radius of the ith circle and N_i is the sum-total number of elements present in the corresponding circle. Here, $r_i = N_i d_i / 2\pi$, d_i is the distance between two antenna in the ith circle. X_{mn} gives the excitation of the current amplitude of jth element present in the ith circle. The value of d_i is set between 0.5λ and λ. θ and ϕ denote the zenith and azimuth angle, respectively, ϕ_{ij} is the central angle of the element of the ith circle, $\phi_{m,n} = 2n\pi/N_i$. Also, θ_0 and ϕ_o gives the direction of the mainlobe and the phase excitation is considered as zero, hence it is set at $\theta_0 = 0$ and $\phi_0 = 0$. Hence, using the above value the equation of field pattern can be simplified as follows (Fig. 7):

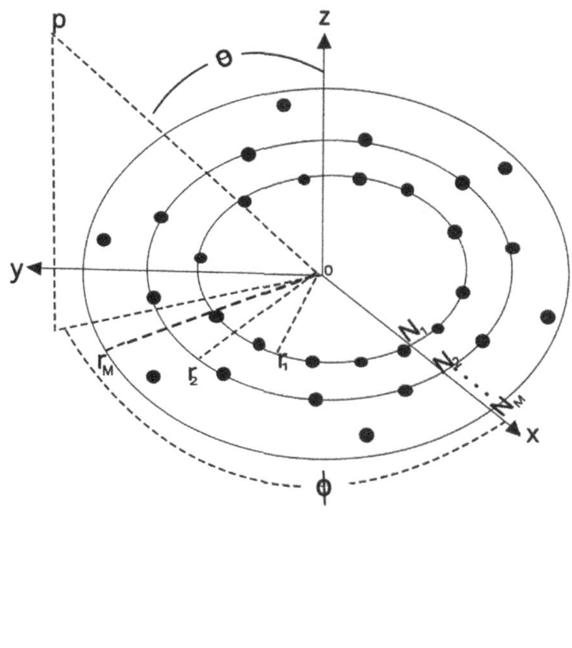

Fig. 7 Geometry of circular concentric antenna arrays

$$E(\theta, \phi) = \sum_{i=1}^{M} \sum_{j=1}^{N_i} X_{mn} \times \exp\{bkr_i \sin(\theta) \cos(\phi - \phi_{ij})\} \quad (20)$$

6.3.2 CCAA Fitness Function

The CCAA's "fitness" function obtained is given by the following formula [7, 8]. Thinning of the CCAA involves obtaining maximal reduction in the value of SLL with minimum usage of elements in the circular rings of the antennae arrays.

$$F = \text{SLLmax} + (T_o^{\text{off}} - T_d^{\text{off}})^2 H(T) \quad (21)$$

Here, SLLmax depicts the maximum value of sidelobe level computed, T_o^{off} and T_d^{off} is the computed and expected number of inactive/turned-off elements, respectively. $H(T)$ depicts the Heaviside Step Function given as follows:

$$H(T) = \begin{cases} 1, \text{ if } T \leq 0 \\ 0, \text{ if } T > 0 \end{cases} \quad (22)$$

$$T = T_o^{\text{off}} - T_d^{\text{off}} \quad (23)$$

6.4 *Experimental Analysis*

The above binSMO is finally applied to check its performance in optimizing the antenna arrays. For the following analysis, a CCAA with two rings with a total of 105 elements is considered. The number of elements in the inner and outer rings is 35 and 70, respectively. The distance between two consecutive elements is $\lambda/2$. The required number of elements to be switched off is to be 43.

For achieving the above-desired output, the binSMO algorithm is run for about 24 times to achieve the optimal results. The various parameter values used in binSMO are set as follows [14] (Table 2).

Following are the experimental results on applying binSMO in contrast with TLBO and FA [14].

On applying binSMO to the two ring CCAA, 43 elements are successfully switched off. Table 3 shows the excitation of all the elements in the rings of CCAA. Here, 1 represents the active elements whereas 0 represent inactive or switched off elements.

Figure 8 given below, represents the same excitation diagrammatically. The following represents the active and inactive elements of the CCAA. Green represents the elements which are active and the rest are switched off.

Table 2 Parameters for the experimental analysis of Binary SMO

Parameters	Values
Perturbation rate	[0.15, 0.45] with linear increase
Population size	100
Maximum no. of spider monkey batches	10
Local leader limit value	[D/2] * swarm size
Global leader limit value	swarm-size/2

Table 3 Element excitation after binSMO is applied

Ring	Elements active and inactive (1 or 0)
1	01111101100100000101110110010110010
2	100010100010001111110001111001000011100011101101111111101000101

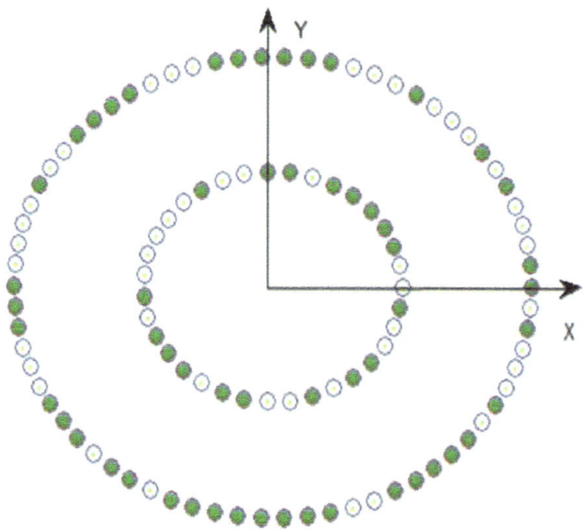

Fig. 8 Diagramatic representation of element excitation of CCAA after binSMO is applied

The experimental results in Table 4 clearly picture SMO having an upper hand with respect to other optimization algorithms. SMO outperformed TBLO and FA by successfully switching off the maximum number of elements. Table 4 shows the results obtained after various algorithms are applied for thinning the CCAA. SMO

Table 4 Results of all the algorithms applied to CCAA

Algorithm	Elements turned-off
SMO	46
TBLO	43
FA	33

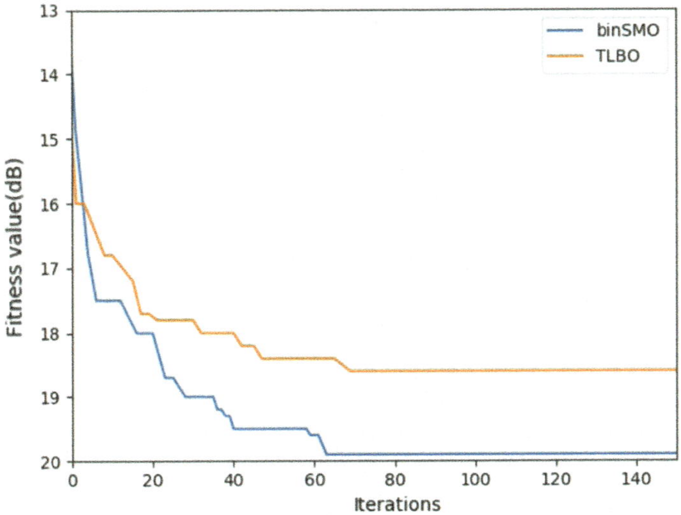

Fig. 9 Fitness value versus no. of iterations curve for binSMO and TLBO applied to CCAA

switched off 46 elements, three greater than the expected value whereas TBLO and FA switched off 43 and 33, respectively. Also, the number of iterations involved in this particular experiment was quite less than that of TBLO and FA (Fig. 9).

The above graph compares the results of binSMO against TBLO. The x axis depicts the number of iterations required to achieve the final result whereas the *y* axis represents the respective fitness values after every iteration. It is evident that the binSMO achieves a better fitness value in less number of iterations as compared to TLBO.

It is observed that the Spider Monkey Optimization performs better in the following cases since it has better exploration–exploitation potential. Also, the results generated by SMO are a result of logical operators XOR, AND and OR, unlike TBLO and FA where the values are rounded-off.

7 Application of Spider Monkey Optimization in Solving the Traveling Salesman Problem

Traveling Salesman Problem (TSP) is a prominent problem in Computer Science. More specifically, TSP is considered as a *combinatorial optimization problem*, and is regarded as a good benchmark for testing the performance of newly developed optimization methods.

> **Note:** An optimization problem *differs* from other algorithmic problems, in that the *better, or most optimal solution* is given more importance than a solution that is computationally more efficient in terms of its speed of execution, memory utilization, etc.

In the following section, the application of Spider Monkey Optimization to solve TSP will be explored and explained.

7.1 Brief Introduction to the Traveling Salesman Problem

The Traveling Salesman Problem (TSP) was proposed by British Mathematician Thomas Kirkman and Irish Math wizard W. R. Hamilton [15]. The TSP description consists of a person (salesman), who needs to travel to "*N*" number of cities (Refer to Fig. 10).

Figure 10 consists of a graph with four nodes that represent the four cities the salesman wishes to visit. Additionally, the cities are linked together by various means of transport, and the cost of traveling along every link has also been depicted in the graph. Let's consider that the salesman starts from city A. The purpose of TSP is to chalk out a path satisfying the following conditions:

- Every city is visited only once
- The cost of travel is to be minimized.

This problem is classified as an NP-Hard problem. Put simply, even if this problem was split into smaller problems, each of these would be just as complex as the original problem.

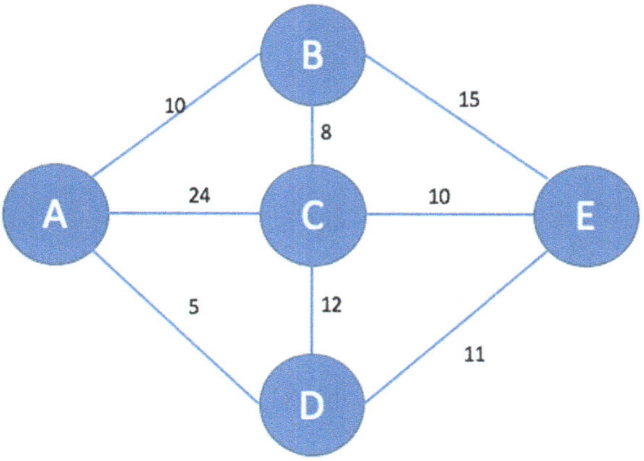

Fig. 10 Sample traveling salesman problem

7.2 *Applying Spider Monkey Optimization to Solve TSP*

The regular Spider Monkey Optimization (SMO) algorithm defined in Sect. 1.2 has been modified in order to solve TSP. Every spider monkey (SM) is considered to be a possible solution to the problem. The modified SMO has six stages, with a few changes made at certain stages in order to solve TSP [16].

Swap Operator in Traveling Salesman Problem

In the next few subsections, the following two terms will be referenced with respect to a TSP solution, namely, SO (*Swap Operator*) and SS (*Swap Sequence*) [17].

For the configuration given in Fig. 8, let's consider the following as a solution:

$$S = (A - B - C - E - D)$$

Now, we wish to execute a Swap Operator (SO) on the solution S, swapping city A with city C. This is denoted by SO(1,3), where 1 and 3 are the indices of the cities that are to be swapped.

$$S' = S + \mathrm{SO}(1, 2)$$
$$S' = (A - B - C - E - D) + \mathrm{SO}(1, 2)$$
$$S' = (C - B - A - E - D)$$

Also, these Swap Operators can be stringed together to form a Swap Sequence (SS).

$$\mathrm{SS} = (\mathrm{SO}(1, 3), \mathrm{SO}(1, 5), \mathrm{SO}(2, 3))$$

Executing a Swap Sequence on a solution

$$S' = S + SS$$
$$S' = (A - B - C - E - D) + \mathrm{SO}(1, 3) + \mathrm{SO}(1, 5) + \mathrm{SO}(2, 3)$$
$$S' = (C - B - A - E - D) + \mathrm{SO}(1, 5) + \mathrm{SO}(2, 3)$$
$$S' = (D - B - A - E - C) + \mathrm{SO}(2, 3)$$
$$S' = (D - B - A - E - C)$$

Additionally, the SS may also be fetched from two solutions. Let's consider them as S_1 and S_2.

The Swap Sequence between S_1 to obtain S_2 can also be obtained using $S_2 - S_1$, as seen in the following equations:

$$\mathrm{SS} = S_2 - S_1 = (\mathrm{SO}_1, \mathrm{SO}_2, \ldots, \mathrm{SO}_n)$$

Let

$$S_1 = (1 - 2 - 3 - 4 - 5)$$
$$S_2 = (2 - 3 - 1 - 5 - 4)$$

Then,

$$E = S_2 - S_1 = \mathrm{SO}(1, 2), \mathrm{SO}(2, 3), \mathrm{SO}(4, 5)$$

7.2.1 Initialization Phase

Just in the same manner as SMO, *N (population size)* random solutions are initialized. Then, to initialize the groups, the population is divided into n groups. LLs in every group is selected. Global Leader is selected from among all of the group members. The following parameters' values have to be initialized in this phase:

- *Maximum_Group (MG) size*
- *GLL-Global_Leader_Limit*
- *LLL-Local_Leader_Limit.*

7.2.2 Local Leader Phase (LLP)

In the LLP, the solution of every member of the local group is modified based on the following:

- Spider monkey's own involvement
- Involvement of members of that group selected at random.
- Involvement of its LL.

The equation for updating the solution for every member of the considered population in this phase is:

$$\mathrm{SS} = U[0 - 1] * (\mathrm{LL}_k - \mathrm{SM}_i) + U[0 - 1] * (\mathrm{RSM}_r - \mathrm{SM}_i) \tag{24}$$

$$\mathrm{BSS} = \text{Basic Swap Sequence}(\mathrm{SS}) \tag{25}$$

$$\mathrm{SMnew}_i = \mathrm{SM}_i + \mathrm{BSS} \tag{26}$$

First, the *Swap Sequences* (SS) between the LL *of the* kth *group* (LL_k) and the ith *member of the population* (SM_i), and between *randomly selected Spider Monkey* (RSM_r) and SM_i are calculated. A random portion of the SS is used from a *randomizer generating numbers* in the range [0, 1] depicted as U[0-1] *in* Eq. (24). Then these two SSs are added to make one SS (using Eq. (24)). *Basic Swap Sequence* (BSS) of the newly generated SS is then calculated using Eq. (25). The BSS thus found is then applied along with the solution of SM_i to obtain new solution of each spider monkey, as depicted in Eq. (26). If the previous solutions' cost is found to be greater than the cost of the new solution, then the new solution replaces the previous solution. Else, no change is made.

7.2.3 Global Leader (GLP) Phase

In the GLP, solutions of all of the spider monkeys in the entire considered population are modified based on the following:

- Spider monkey's own involvement
- Involvement of local group members selected at random
- Involvement of the global leader.

Hence, we can see that this phase is very similar to the Local Leader Phase discussed in Sect. 1.2.2. However, in this phase, every SM decides whether it will update its solution based on a probabilistic value (Refer Eq. (27) for the formula used to calculate the probability of updating SM_i)

$$\text{prob}(i) = 0.9 * \frac{\text{minimun} - \text{cost}}{\text{cost}(j)} + 0.1 \tag{27}$$

In Eq. (27),

Parameter	Description
• cost(*j*)	Touring cost of spider monkey
• minimum_cost	A group of spider monkeys' least tour cost

Based on the probability value prob(i) obtained, the results are updated in this phase based on the following equations:

$$\text{SS} = U[0, 1] * (\text{GL} - \text{SM}_i) + U[0, 1] * (\text{RSM}_r - \text{SM}_i) \tag{28}$$

$$\text{BSS} = \text{Basic Swap Sequence(SS)} \tag{29}$$

$$\text{SMnew}_i = \text{SM}_i + \text{BSS} \tag{30}$$

These three equations are very similar to Local Leader Phase equations. They function in the same manner as in the LL-Phase, except for the probability-based updation of solution (using Eq. (27)). A graphical representation of the solution updation process is depicted (Refer to Fig. 11).

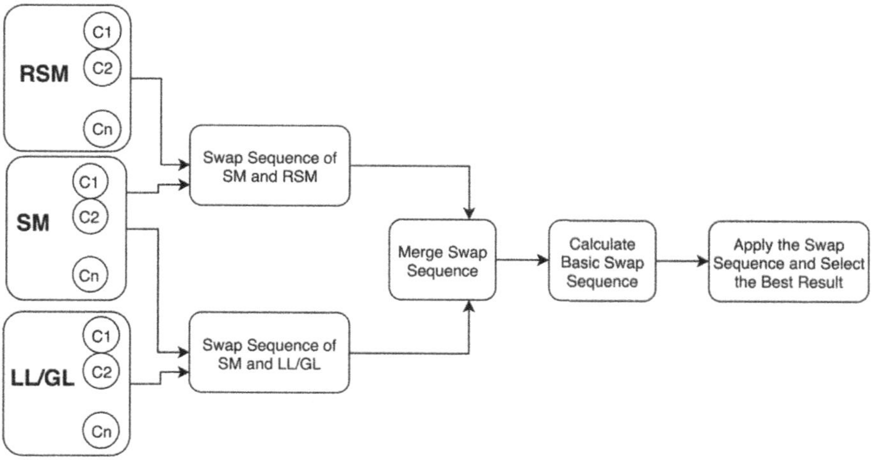

Fig. 11 Solution updation process

7.2.4 Local Leader Learning (LLL) Phase

The LLL-phase involves comparison of the cost of touring for the LL in every group and the minimum cost of touring. Minimum cost of touring is compared with the LL's cost of touring.

- If the LL cost is found to be greater, then the new SM solution is chosen as the group's new LL
- Else, the current solution of the LL is retained.

The LLC-LocalLimitCount value is stepped up by 1 if the LL is not modified.

7.2.5 Global Leader Learning (GLL) Phase

In the GLL phase, calculating the minimum tour of all the SM, and comparing it with the current tour cost of the global leader.

- If the current tour cost of the GL is found to be greater, then the new SM solution is used to update the GL. Additionally, when the cost of GL is unchanged, the GlobalLimitCount is stepped up by 1.

7.2.6 Local Leader Decision (LLD) Phase

If for a defined time-period *(Local_Leader_Limit)*, the position of LL is not changed, then every member of that group modifies their position by using either of the following two methods:

- Employing random initialization of positions
- Using combined participation of LL and GL (as seen in Eq. (31)) to determine positions

$$\text{SMnew}_i = \text{SM}_i + U[0, 1] * (\text{GL} - \text{SM}_i) + U[0, 1] * (\text{SM}_i - \text{LL}_k) \quad (31)$$

Here, GL denotes the global leader while the kth group's GL is LL_k. Following the Eq. (31), SM_i detaches from LL and moves forward to the GL.

7.2.7 Global Leader Decision Phase (GLD)

If for a defined time-period *(Global_Leader_Limit)*, the location of the global leader is not modified, then multiple small groups are formed by the GL-Global Leader by splitting the population until *Maximum Group Size* (MG) is reached. In this phase, every time the group is split, the new LL is chosen by the previously defined LLL-phase (discussed in Sect. 7.2.4). When the number of created groups exceeds the value of MG, the global leader merges all of these smaller groups back into a single group.

7.3 Algorithm Definition and Worked Example

To better understand the modified SMO algorithm to solve TSP, consider the following example [24] (Refer Fig. 12). After completing the initialization phase, the

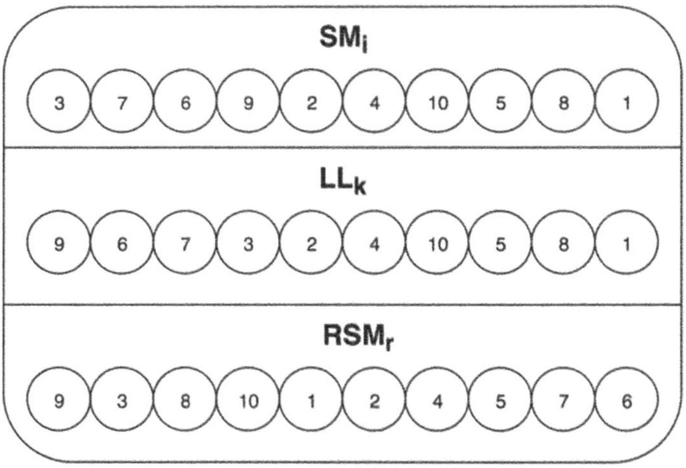

Fig. 12 After initialization phase

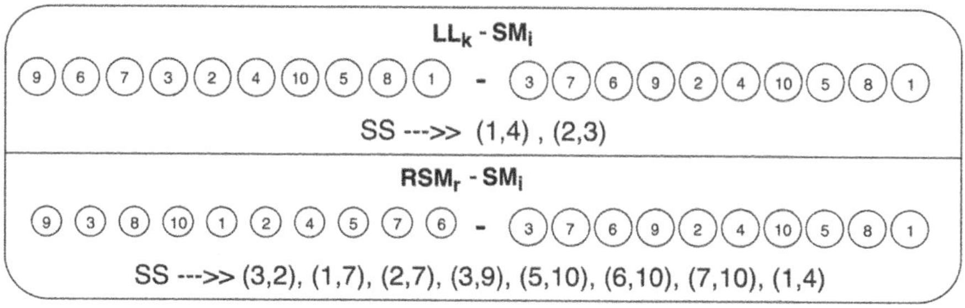

SS = U[0-1] * (LL_k – SM_i) + U[0-1] * (RSM_k – SM_i)

SS = U[0 - 1] * {(1,4), (2,3)} + U[0 - 1] * {(3,2), (1,7), (2,7), (3,9), (5,10), (6,10), (7,10), (1,4)}

Suppose, U[0 – 1] = 0.5, and U[0 – 1] = 0.3.
SS = (0.5 * {(1,4),(2,3)} + 0.3 * {(3,2), (1,7), (2,7), (3,9), (5,10), (6,10), (7,10), (1,4)})
SS = (2,3), (3,2), (1,7), (2,7), (3,9), (1,4)

BSS = BasicSwapSequence (SS)
= BasicSwapSequence ((3,2), (2,3), (1,7), (2,7), (3,9), (1,4))
= (1,7), (2,7), (3,9), (1,4)

Fig. 13 Finding SS and BSS (Local Leader Phase)

values of the *population*, *local leaders*, and *randomly selected spider monkeys* can be seen in Fig. 12.

Now, proceeding to the LL-Phase, Eq. (1) is used to find the swap sequence SS and Basic Swap Sequence BSS (Refer to Fig. 13).

The BSS, along with the existing SM solutions are applied together to find the new SM solutions (Refer Fig. 14).

8 Research Trends and Conclusion

Over the past few years, Nature-Inspired Computing has been used to devise innovative and effective solutions to various optimization problems. The Spider Monkey Optimization (SMO), described in this chapter, is one such nature-inspired algorithm developed as recently as 2014 by Bansal et al. [1]. This chapter discusses the methodology, types, and applications of SMO, with focus given on implementations of SMO to solve real-world problems with case studies such as optimally using the resources of a dam and scheduling the dam operations. Variants of SMO with specific applications in data science [29] have also been explained. One such example is the Constrained SMO algorithm, which has been shown to be an able candidate for the task of global optimization of constrained optimization problems. The chapter highlights the use of SMO to resolve the classic Traveling Salesman Problem (TSP), which is the de facto combinatorial optimization problem used to

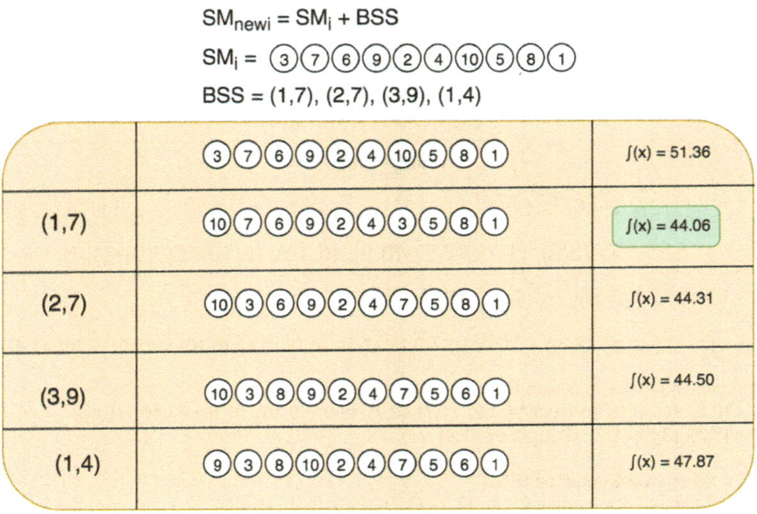

Modified SMO Algorithm to solve TSP

```
1: Initialisation
2: Choose GL and LLk of the kth group
3:    For every solution SMi in the group
4:          Eq.(24) to find SS, Eq.(25) to find BSS, Eq.(26) to change value of SMi
5:          Eq.(27) to calculate the probabilistic value prob(i)
6:          Eq.(28) to find SS, Eq.(29)to find BSS, Eq.(30) to change value of SMi
            using prob(i) to decide when to update SMi.
7:          Update LLk and GL
8:          Initialise member positions by either employing random
            initialisation of all of the members or use Eq. (31)
9:          Splitting of population into smaller groups and/or combine all the
            smaller groups
10:         Check termination condition, if met, then proceed to Step 3.
11:The obtained solution is that of the Global Leader
```

Fig. 14 Finding new SM solutions (Local Leader Phase)

test new optimization algorithms. The results obtained, which are very promising proving the credibility of SMO, are showcased in this chapter as a case study.

Spider Monkey Optimization is a recent five-year-old algorithm. Research trends show work being done in performance analysis of SMO with other probabilistic and deterministic algorithms. Some variants of SMO have been developed which uses different perturbation rates, hybridization, and self-adaptive strategies. Optimization is a fertile area of research, meaning that various such variants of SMO are sure to emerge in the near future.

References

1. Bansal J, Sharma H, Jadon S, Clerc M (2014) Spider Monkey Optimization algorithm for numerical optimization. J Memetic Comput 6. https://doi.org/10.1007/s12293-013-0128-0
2. Ehteram M, Karami H, Farzin S (2018) Reducing irrigation deficiencies based optimizing model for multi-reservoir systems utilizing spider monkey algorithm. Water Resour Manag 32:2315–2334
3. Kumar Sandeep, Sharma Vivek, Kumari Rajani (2015) Self-adaptive spider monkey optimization algorithm for engineering optimization problems. Int J Inf Commun Comput Technol (IJICCT) II:96–107
4. Kumar S, Sharma B, Sharma VK et al (2018) Evol Intel. https://doi.org/10.1007/s12065-018-0186-9
5. Agrawal A, Farswan P, Agrawal V, Tiwari D, Bansal J (2017) On the hybridization of spider monkey optimization and genetic algorithms
6. Radclie NJ (1991) Equivalence class analysis of genetic algorithms. Complex Syst 5(2):183–205
7. Michalewics Z (1996) Genetic algorithms data structures of evolution programs
8. Schlierkamp-Voosen D (1994) Strategy adaptation by competition. In: Proceedings of the second European congress on intelligent techniques and soft computing, pp 1270–1274
9. Haupt L (1997) Phase-only adaptive nulling with a genetic algorithm. IEEE Trans Antennas Propag 45(6):100915
10. Liao WP, Chu FL (1999) Array pattern synthesis with null steering using genetic algorithms by controlling only the current amplitudes. Int J Electron 86(4):44557
11. Khodier M, Al-Aqeel M (2009) Linear and circular array optimization: a study using particle swarm intelligence. Prog Electromagn Res 15:34773
12. Khodier MM, Christodoulou CG (2005) Linear array geometry synthesis with minimum sidelobe level and null control using particle swarm optimization. IEEE Trans Antennas Propag 53(8):26749
13. Dib N, Sharaqa A (2014) Synthesis of thinned concentric circular antenna arrays using teaching learning based optimization. Int J RF Microw Comput Aided Eng 24(24):44350
14. Singh U et al, A novel binary spider monkey optimization algorithm for thinning of concentric circular antenna arrays
15. The traveling salesman problem. https://www.csd.uoc.gr/~hy583/papers/ch11.pdf. Accessed 25 Apr 2019
16. Ayon SI, Akhand MAH, Shahriyar SA, Siddique N (2019) Spider monkey optimization to solve traveling salesman problem. In: 2019 international conference on electrical, computer and communication engineering (ECCE), 7–9 February 2019
17. Soni N, Kumar T (2014) Study of various mutation operators in genetic algorithms. Int J Comput Sci Inf Technol (IJCSIT) 5(3):4519–4521

18. Sharma A, Sharma A, Panigrahi BK, Kiran D, Kumar R (2016) Ageist spider monkey optimization. https://doi.org/10.1016/j.swevo.2016.01.002 (Received 15 Jul 2015, Revised 12 Dec 2015, Accepted 18 Jan 2016)
19. Agrawal V, Rastogi R, Tiwari DC (2018) Spider monkey optimization: a survey. Int J Syst Assur Eng Manag 9(4):929–941. https://doi.org/10.1007/s13198-017-0685-6
20. Kumar S, Sharma B, Sharma VK, Poonia RC (2018) Automated soil prediction using bag-of-features and chaotic spider monkey optimization algorithm (Special Issue, VK et al). Evol Intel. https://doi.org/10.1007/s12065-018-0186-9
21. Hazrati G, Sharma H, Sharma N, Bansal JC (2016) Modified spider monkey optimization. https://doi.org/10.1109/iwci.2016.7860367
22. Gupta K, Deep K (2017) Investigation of suitable perturbation rate scheme for spider monkey optimization algorithm
23. Singh D, Salgotra R, Singh U (2017) A novel modified bat algorithm for global optimization. https://doi.org/10.1109/ICIIECS.2017.8275904. 17–18 March 2017
24. SMO to solve TSP. https://github.com/SafialIslam302/Thesis—Modified-SMO-to-Solve-TSP/blob/master/1407041%20-%20Modified%20SMO%20to%20Solve%20TSP.pptx. Accessed 25 Apr 2019
25. Das H, Jena AK, Nayak J, Naik B, Behera HS (2015) A novel PSO based back propagation learning-MLP (PSO-BP-MLP) for classification. In: Computational intelligence in data mining, vol 2. Springer, New Delhi, pp 461–471
26. Rout M, Jena AK, Rout JK, Das H (2020) Teaching–learning optimization based cascaded low-complexity neural network model for exchange rates forecasting. In: Smart intelligent computing and applications. Springer, Singapore, pp 635–645
27. Sneha V, Shrinidhi KR, Sunitha RS, Nair MK (2019) Collaborative filtering based recommender system using regression and grey wolf optimization algorithm for sparse data. In: 2019 International conference on communication and electronics systems (ICCES) (in press)
28. Nayak J, Naik B, Jena AK, Barik RK, Das H (2018) Nature inspired optimizations in cloud computing: applications and challenges. In: Cloud computing for optimization: foundations, applications, and challenges. Springer, Cham, pp 1–26
29. Sahani R, Rout C, Badajena JC, Jena AK, Das H (2018) Classification of intrusion detection using data mining techniques. In: Progress in computing, analytics and networking. Springer, Singapore, pp 753–764

Chapter 7
Multi-agent-Based Systems in Machine Learning and Its Practical Case Studies

K. R. Shrinidhi, Sneha V, Vybhav Jain, and Mydhili K. Nair

Abstract In the area of Decision Science research, Multi-Agent Systems (MAS) has been hailed as a new paradigm for conceiving and designing software systems. The goal of this chapter is to underscore the need for and usefulness of MAS by giving the reader an insight into the agents' characteristics, its interaction with its environments, various performance measures and different types of MAS. This chapter gives insights on Reinforcement Learning Agents the root of which originates from animal learning theory. In reinforcement learning, the agents comprehend the state of the environment at each step, and adopt the decision on the action to be taken, causing its environment to transit into a new state. The agents each receive a scalar reward that evaluates if the transition made is aligned with the goal. The novelty of this chapter is the use of a multi-agent programmable tool called *NetLogo* to explain the reinforcement learning technique with appropriate examples and the results obtained are thoroughly analyzed. Prominent algorithms like Policy Hill-Climbing, Q-learning, and its variants such as Nash Q, Minimax Q-Learning, and optimization techniques such as Distributed Constraint Optimization (DCOP) in multi-agent systems which help the agents in achieving equilibrium are discussed. The chapter also focuses on the typical domains that can benefit from MAS. It comprises of instances from examples of MAS implemented using the NetLogo models for enhancing the understanding of agents and their environment. The chapter concludes by giving the reader a sneak-peek into the latest research trends in MAS.

Keywords Multi-agent-based systems · Reinforcement learning · Equilibrium algorithms · Netlogo

---

K. R. Shrinidhi · S. V · V. Jain · M. K. Nair (✉)
Department of Information Science and Engineering, Ramaiah Institute of Technology, Bangalore 560054, India
e-mail: mydhili.nair@gmail.com

J. K. Rout et al. (eds.), *Machine Learning for Intelligent Decision Science*,
Algorithms for Intelligent Systems, https://doi.org/10.1007/978-981-15-3689-2_7

1 Intelligent Agents and Its Interacting Environment

A software agent is a system that has the ability to act upon the information that it has perceived. Intelligent agents are agents that are capable of taking actions independently so as to satisfy the required objectives by interacting with their environment and using past experiences. These agents have the ability to perceive information from their environment through sensors and they use this information to improve their decisions that help in achieving the objectives. They are called intelligent agents because they are able-bodied to adapt to their changing real-time environment and are able to analyze themselves with respect to their past behavior, success and error rate. Generally, the intelligent agents can be designed to autonomously gather information of interest from its environment on a regular basis, predefined schedule or when requested by the user in real-time. The user input is collected by the intelligent agents using sensors, like microphones or cameras, and agent output is delivered through actuators, like speakers or screens. Based on the input collected, the agents decide their moves to reach desired goals.

Examples

1. Humans are viewed as agents as they respond to the perceived information from their surroundings through their sensory organs.
2. In the ecosystem, predators and preys are also regarded as agents as they have objectives to survive which are achieved by being vigilant of intruders in their surrounding environment (discussed further in Sects. 1.1 and 2.4).
3. Robots like COG from MIT [1], Xavier from CMU [2], Mars Lander (discussed in Sect. 3.4), and AIBO entertainment from Sony [3] are agents with sensors to receive information and have the capability to take decisions and actions.

1.1 Illustrating the Example of a Predator–Prey Model

In order to understand intelligent agents an example of Wolf (Predator) and Rabbit (Prey), two intelligent agents that take actions after perceiving the various factors in their environment are explained in Table 1.

1.2 Agent Function

The **agent function** can mathematically now be defined as a function f(x) which maps every potential percepted information sequences obtained from environment to a potential action that the intelligent agent can execute which then affects the forthcoming actions.

Table 1 Example of predator and prey (agents) taking actions on sensing their environment. It is to be noted that here the prey is in the environment of the predator and vice versa

Intelligent agent	Objective	Information perceived from environment	Action taken by the intelligent agent
Wolf (Predator)	Satisfying its hunger	• Senses the sound and smell of prey in its surrounding	• Waits for the prey to be vulnerable, prepares itself for the attack by increasing metabolism
		• Observes the movements of the prey and identifies its vulnerability	• Uses various strategies like stealth and speed to attack the prey
Rabbit (Prey)	Survival	• Through its sensory receptors, it perceives the presence of predator by position, sound, and sight	• Prepares itself for escape and increases the rate of metabolism likewise in order for its survival
			• Uses various strategies to deceive the predator for its survival

```
f (percepted information sequence) -> potential action to be taken
```

- For the wolf, at an instance, the agent function f can be defined as
  `f (identifies the vulnerable moment of prey) -> attacks the prey`
- For the rabbit, at an instance, the agent function f can be defined as
  `f (senses the presence of an intruder) -> prepares itself for escape`

1.3 Discussing the Environments of the Agents

Since the agent perceives its environment and uses the data it collects to make informed decisions, the agent-environment is a vital factor in designing intelligent agent systems. The classification of the types of environment the agents operate on is given below in Table 2.

Table 2 Classification of the types of environment the agents operate on

1	Deterministic environment versus stochastic environment	
	When an agent executes an action with no uncertainty about the outcome or consequences (state of the environment), it is a deterministic environment. **Example**—A game of Tic-tac-toe where if a player scores three cells either horizontal, vertical or diagonal, the player is bound to win and this outcome does not change	A stochastic environment is one in which the agent does not uniquely determine a state it will transition to. This could be because of the element of probability that creeps in due to the lack of complete sensor coverage or due to an inherent need for randomness within the system. **Example**—Mars Lander robot (discussed in Sect. 3.4)
2	*Completely accessible versus incomplete access environment*	
	Completely accessible environments are those environments where at any given instance of time, agents possess up-to-date information about all the environment's state to complete a branch of the problem. **Example**—A game of computer vs human chess where both players are aware of all the states in the game	If the agent cannot anticipate the move toward accurate state in advance and has to focus on finding an equilibrium at any given instance of time due to noisy, inaccurate or missing sensors, it is an incomplete environment. **Example**—A game of bridge with cards where the move of players are not anticipated
3	*Discrete environment versus Continuous environment*	
	If there are a finite number of states to reach the final state (goal state), it is a discrete environment. **Example**—A game of Sudoku or Tic-tac-toe where the grid size of the game is fixed	If the states in the environment are constantly being updated with time, it is a continuous environment. **Example**—A game of Ludo or Snake and Ladder where each time the player is punished, the number of states to reach the end of the game increases
4	*Static environment versus dynamic environment*	
	If the environment does not change until and unless an action is taken by the agent, it is a static environment. **Example**—A human deciding which restaurant to go to for dinner or which car to purchase	If the environment changes without an agent controlling and taking decisions, it is a dynamic environment. In a dynamic environment, the objective functions of the system may or may not change. **Example**—A game of snake and ladder with moves decided by the call of dice
5	*Episodic environment versus sequential environment*	
	Episodic environments are environments where agents depend only on the current state of memory to make decisions. **Example**—Touch-me-not plant and simple driver system (discussed in Sect. 3.1)	Sequential environments require considering memory of past actions taken by agents to make decisions on the next best action. **Example**—Poker, Soccer, Tennis, Cricket, Backgammon etc.

(continued)

Table 2 (continued)

6	*Competitive environment versus collaborative environment*	
	If the agents are working against each other, where each agent is trying to be the best agent, it is a competitive environment. **Example**—Arcade games like Pubg and CounterStrike with human vs AI bots playing against each other	If all the agents in the environment are working together to achieve the common objective of the system, it is a collaborative or cooperative environment. **Example**—Supply chain management agents and Ant colony based agents (discussed in Sects. 7.1 and 7.2)

2 The Paradigm of Multi-Agent Systems

Nature has evolved over a billion years and is a rich source of inspiration. Contemplate any ecosystem, several self-ruled independent species are noticed to cohabitate together with interdependencies and complex interactions. The analogy between Multi-Agent Systems (MAS) and ecosystems is self-evident. In fact, every entity in the ecosystem can be considered as an agent [4]. Therefore, it is efficient to model them with agents. Animal species foraging food is one of the best examples that can be given since they interact with each other and its surrounding environment to attain the objective of finding food. Spider-monkeys are an endangered species possessing great threat from their predators, hence it becomes important for them to have a strategic pattern for foraging. During the day, they form loose groups of 10–20 individuals who avoid each other depending on the competition of food and risk of predation. At the end of the day, each group strives to return home safely with their food. Each monkey here is analogous to an "intelligent agent" that interacts with other agents (monkeys) in the group and are vigilant to its surrounding environment with the primary objective of all "agents" in this "multi-agent spider monkey system" is being safe and successful food-foraging. To support this analogy, various nature-inspired optimization algorithms have seen the face of applications in the domains of recommender systems [5], cloud computing [6], data mining [7], big data analytics [8, 9], and so on [10]. Refer the figure which explains the analogy and shows the interactions between the agents and environment Fig. 1.

In this epoch of diversified technologies, inspired from nature, Multi-Agent Systems (MAS) have hailed as a new paradigm for conceptualizing, designing, and implementing software systems. MAS is a collection of multiple intelligent agents that interact with each other and their environment in order to achieve the final objective of the system. These are sophisticated systems that act autonomously across distributed and open environments, to solve complex problems that cannot be easily comprehended by an individual agent or a human.

In the digital age today, computers are no longer the traditional stand-alone systems. Computers are now tightly coupled with the users as well as one another. Modern computing environments are heterogeneous, widely distributed, large, and open in nature. To cope with such complex applications built in these environments, computers have to behave more like "individuals" or "agents", rather than acting just

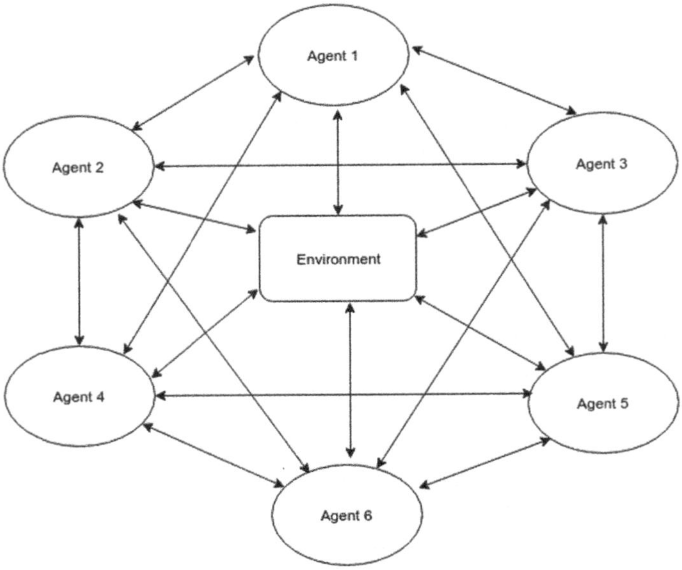

Fig. 1 Interactions between multi-agents and its environment

as "parts." The tasks need to be distributed between the agents and after analyzing the dependencies between these agents, these tasks need to be executed parallelly as shown in Fig. 1. For example, applications like supply chain management, financial portfolio management, airplane maintenance, and military logistics planning require multiple agents that can work parallelly together to solve complex problems efficiently and to optimize the time taken to achieve the set targets of the system.

2.1 Characteristics of Multi-Agent Systems

The prominent characteristics of agents constituting the Multi-Agent Systems are as follows [4]:

- **Self-rule**: Each agent has its own set of goals to achieve and are allotted the required resources for its functioning. Each agent can take their decisions autonomously without any need for guidance from other agents. Further, the agents are not to be authorized to order actions to be taken or dictate decisions to other agents in the system. This contributes to system robustness through modularity. In addition, identifying and resolving faults in the system becomes easy. When an agent is found to contribute error, that particular agent is isolated from the system and the spread of error can be avoided.
- **Interaction**: Includes the ability of cooperating, coordinating, and negotiating between agents in the system.

- **Local views**: No agent possesses the complete knowledge or the global view of the system since the actions of other agents are not pre-known. Since real-time systems are intricate in nature, providing the global view of the system over-burdens the agent, and can lead to exploitation of knowledge by agents.
- **Decentralization**: As discussed in the need for MAS above, the task needs to be distributed amidst all the agents in the system so as to provide flexibility and to optimize the time taken to achieve the end goal of the system.
- **Communication**: The agents can communicate asynchronously. They can also work parallelly.

2.2 Parameters Associated with Multi-Agent Systems

Since multiple intelligent agents form a complex network of interaction in the system, these systems necessitate a lot of investigation and scrutiny for maintaining good performance measures such as number of agents, time taken for execution, etc. [11]. The various parameters contributing to the performance of the MAS are shown in Table 3.

Table 3 Parameters contributing to the performance of the multi-agent systems

S. no	Parameter	Relation with performance of the system
1	Number of agents	As the number of agents increases in the system, the system tends to grow complex due to the multiple interactions. Therefore the complexity of a MAS is directly proportional to the count and type of agents that constitute the MAS
2	Time taken for computation	An agent is said to be in an active state whenever it is executing its designated tasks. The computational time for an agent is hence calculated as the total time for which an agent is in the active state
3	Status of the agent	Each agent executes an action that results in a state. If the switches in state are high, performance is said to be poor
4	Coordination between agents	The tasks are distributed among the agents which require coordination between the agents to accomplish the goal. When there is no proper communication channel established between the agents, the system is highly prone to risks and error. If the cooperation between the agents is achievable, then the performance of the system is assured

2.3 Quick Introduction to Netlogo—A Multi-agent Programmable Modeling Environment

NetLogo is a Multi-agent programmable modeling language [12]. It is one among the standard platform for agent-based simulation used by researchers and students for modeling the agents and its interacting environment. The primary domain of NetLogo is in the fields of social and natural sciences. Netlogo is a free open source software under the GPL which has a user-friendly environment to learn and hone up your multi-agent-based modeling skills. After StarLogo and StarLogoT, NetLogo is the next generation of multi-agent modeling languages [12]. NetLogo runs on the Java Virtual Machine (JVM), therefore it works on all major platforms (Mac, Windows, Linux, etc.). It also has a number of libraries containing various examples and simulations including some nature-inspired multi-agent models using reinforcement learning like Ants Foraging, Building Honeycombs, Wolf Sheep Predation being a few of the examples under biology, one of the many domains covered by NetLogo. Furthermore, NetLogo allows authors modeling of new multi-agent systems and authorizes altering and transforming the existing models according to user requirements.

Downloading and installation of NetLogo is very simple [13]. The agent-based models can be executed and viewed on their GUI and different parameters assigned to the agents or the environment can be altered to see its effect on the agent and its interacting environment. This chapter comprises of the various case studies and results obtained by the agents for a few sections, demonstrated using NetLogo for enhancing the understanding of our readers.

2.4 Demonstrating Agents and Their Environment Using Netlogo

The Predator–Prey model discussed below explores the stability of ecosystems as discussed in Sect. 1. This model demonstrates the agents and their interaction with the environment using NetLogo. An ecosystem is said to be stable if it prevents the extinction of species involved. Despite fluctuations in population sizes (number of agents), a system must always be stable. Two kinds of agents are in the model—the sheep and the wolf that wander randomly in the environment, where the wolves hunt for the sheep to feed on while the rabbit eats the grass. Each move costs the wolves some amount of their energy, and they must eat the sheep in order to replenish their energy to avoid death by running out of energy. Similarly the sheep must graze the grass to replenish their energy. Clearly, the objectives of the agents are to avoid the predator, consume food and survive and this is a good example of a nature-inspired goal-based agent system discussed later in Sect. 3.3. To allow the growth of the population of both species to maintain a balance, each wolf or sheep has a preset probability of reproducing at each time step. For executing the model, the parameters like number of sheep and wolves, the energy replenished by the sheep and wolf on eating their food and the production rate of the grass are set and are shown in Fig. 2. In Fig. 3, we can see brown patches which indicate the grass eaten by the sheep and it is noticed the sheep and wolf count have increased by 56 and 21, respectively. The graph of population depicted in Fig. 3 shows the growth of population of wolves,

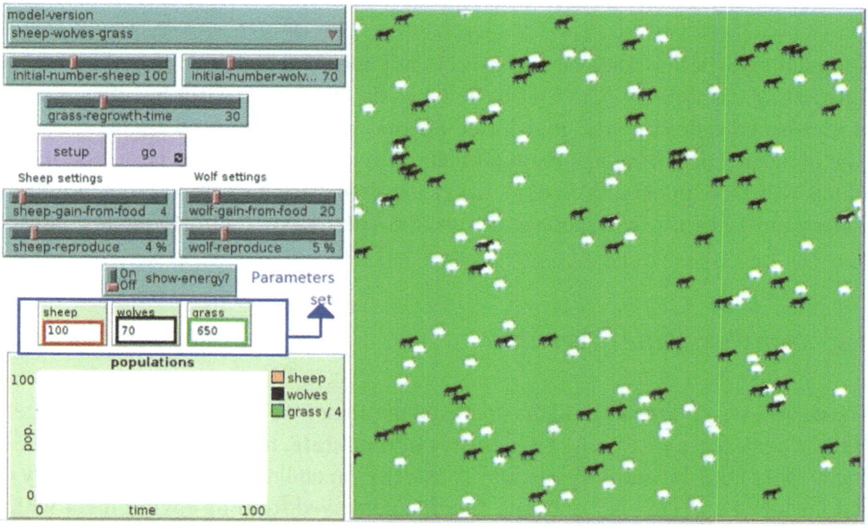

Fig. 2 Parameters set for the agents and its environment

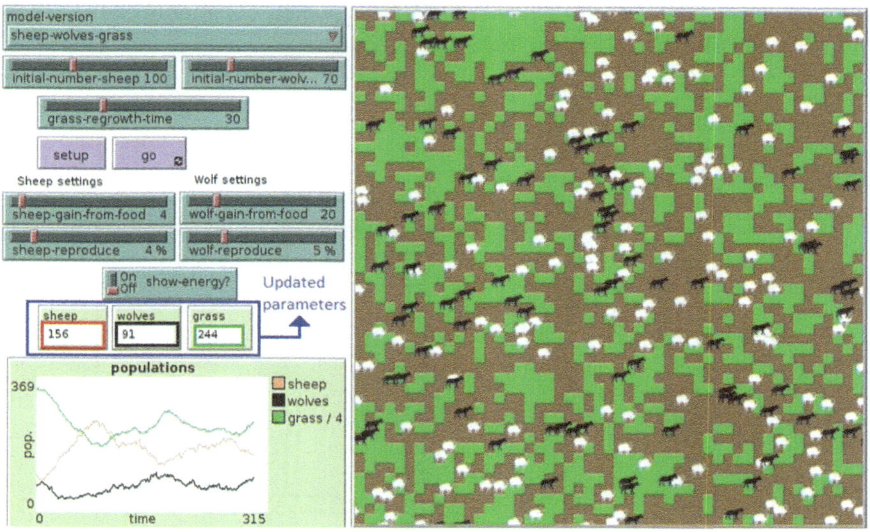

Fig. 3 Predator–Prey model execution

sheep, and grass with time. Readers can change the parameter value and take a look at the effects in the model [14].

3 Types of Agents

The agents can be classified on the basis of their degree of perceiving intelligence and functioning capabilities. The four basic types of agents are—**simple-reflex agents, model-based reflex agents, goal-based agents, and utility-based agents**. The fifth type of agent is **learning agents** that functions by learning from its past experiences which will be explained in Sect. 4.1. Let us discuss each of these types briefly to understand the behavior of these agents and the suitable environment for these agents.

3.1 Simplex-Reflex Agents

The easiest type of agents to understand and implement is the Simplex-Reflex Agent with its central premise being "**if *x* is the current state, then take action *y***". These agents only take the current state into consideration and neglects both the history of actions taken and the history of information perceived from its environment which leads to the agent taking actions randomly. The type of environment suitable for simple-reflex agents is episodic and completely accessible demonstrating actions based on current situational experiences. The behavior of these agents as depicted in Fig. 4 can be related to the behavior of Mimosa pudica, commonly called Touch-Me-Not plant that shrivels as a reflex of being touched.

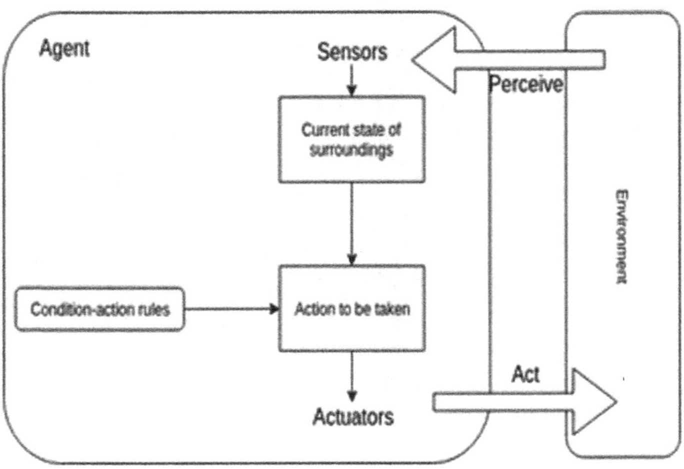

Fig. 4 Simplex-reflex agent

```
function SIMPLE-REFLEX-AGENT(precept) returns action
    static: rules, a set of condition-action rules
    state -> INTERPRET-INPUT(precept)
    rule -> RULE-MATCH(state,rules)
    action -> RULE-ACTION[rule]
    return action
```

Fig. 5 Algorithm for simple-reflex agents

The algorithm shown in Fig. 5 begins with initializing the set of "`if state->then action`" rules for the agent. For a basic remote control car, a set of rules like "`If the obstacle is encountered, then turn on beeper`" and "`If input signal is to go right, then rotate wheels to right and move right`" needs to be initialized. For the percept given, function "`INTERPRET-INPUT`" produces the abstract or the description of the agent's current state. The function "`RULE-MATCH`" returns the first matched rule in the set of predefined rules that match with the obtained description of the agent's state. "`RULE-ACTION`" function then implements the decision according to the matched rule. The environment is clearly "episodic" and actions are not dependant on the previous actions.

3.2 *Model-Based Reflex Agents*

Unlike the simple agents, Model-based agent is able to store its precept history or sensed information in its knowledge-based or memory, which helps it to make informed decisions. It also considers the fact that its surrounding will change and this will affect its actions on the surrounding. It acts based on the condition-action rules. The current state of the agent is saved in a structured format which depicts the parts of the environment that is invisible to it. This type of agent requires maintaining the knowledge of how its surrounding work and evolve, this "knowledge" is known as a "model of the environment" and therefore the name "model-based reflex agents". The agents must maintain an internal model which provides information regarding the percept history that can help in understanding at least some unknown parts of the current state. Research-based planetary system can predict the planet's climatic conditions based on geo-information knowledge base which is a good example of where Model-based Reflex Agents can be implemented as shown in Figs. 6 and 7.

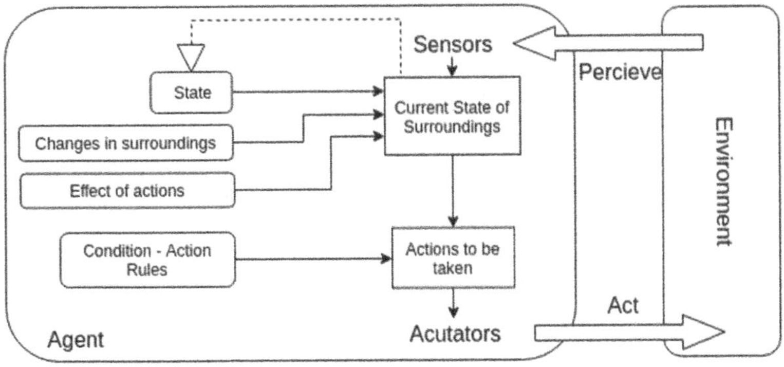

Fig. 6 Model-based reflex agent

```
function MODEL-BASED-REFLEX-AGENT(percept) returns action
    static: state, a description of the current environment state
            rules, a set of condition-action rules
    state -> UPDATE-STATE(state,precept)
    rule -> RULE-MATCH(state,rules)
    action -> RULE-ACTION[rule]
    state -> UPDATE-STATE(state,action)
    return action
```

Fig. 7 Algorithm for model-based reflex agents

3.3 Goal-Based Agents

All living organisms have a common goal, to survive. Let it be humans, animals or plants, we all make decisions according to our environment in order to meet this goal. A good example to depict this type of agent can be seen in nature, i.e., the black peppered moth. These agents were originally a mixture of white and black color. But in the early nineteenth century, as the industrial revolution was at its peak, white and black patched moths were unable to hide on trees and this resulted in their reduced numbers. To counter this, every following generation of moths could be seen to have lesser and lesser white patches thus enabling a population with mostly or all black peppered moths. Thus, the moths are now able to survive longer.

Building upon the concept of model-based reflex agents discussed in Sect. 3.2, goal-based agents depend upon "goal information" rather than "condition-action rule" as in model-based reflex agents. The desired final outcome state that the agent has to achieve ultimately is the "goal" of the agent. To give an example of a goal-based agent, consider the realm of Data Mining. To give a brief insight, Data Mining is the process of garnering information from huge databases that was previously incomprehensible or unknown and then transforming it into understandable structures for making decisions. Here multi-agent systems inspired from the ant colony are "goal-based agents" whose goal is to discover the classification rules. These agents are not

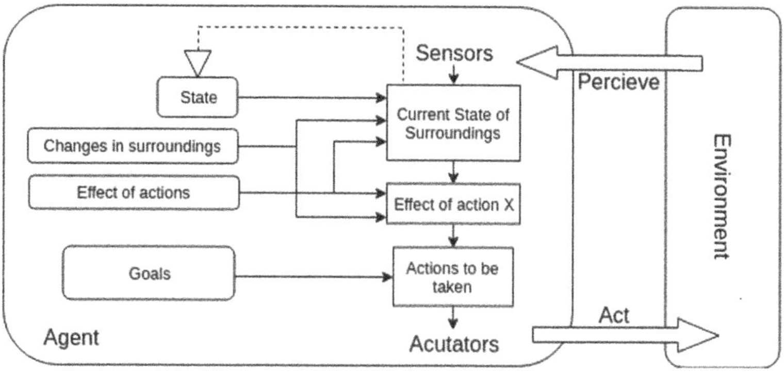

Fig. 8 Goal-based agent

just given a set of IF-> THEN rules. They are given the final objective of predicting the classes for the training set accurately that best fits the data by identifying patterns in them and extracting the classification rules from the data themselves [15]. The actions of a goal-based agent is always intended toward achieving the objective or goal, that is its actions are goal-driven which is depicted in Fig. 8.

The goal information gives the desirable situations and provides a wide range of choices for taking a decision and the agent to choose the best decision that fits in achieving the overall goal and takes the agent closer to the goal state. Effect of each action is considered as is evaluated against the desired goal to with an objective to minimize the distance from goal state. Searching and planning are two important steps for a goal-based agent to achieve their objective. Finally, if the agent performs the action which helps it to achieve its goal, then it is said to be at a "happy" or "desirable" state in which the agent can exist to perform at its best.

3.4 Utility-Based Agents

In goal-based agents, the only concern is about the goal state which distinguishes between "happy" and "unhappy" states, in other words, can be termed as "desirable" and "undesirable" states. But it is also essential to highlight how much desirable a particular state is and measure the "happiness" of the agent. For example, though it is essential to look for efficient and quicker ways to complete a task, it is equally important to consider how "happy" and agent is in a particular state can be scientifically determined by the "utility" of that state. This measure of "happiness" of the agent is assigned by the utility function, which maps a state to the degree of its utility (happiness). Now, the agent opts for choosing the state maximum measure of utility, thereby having efficient and optimal ways of reaching the final objective. In Fig. 9, it is observed that the agent takes into the fact the feasibility and utility of the state along with the effect of its action in the surrounding.

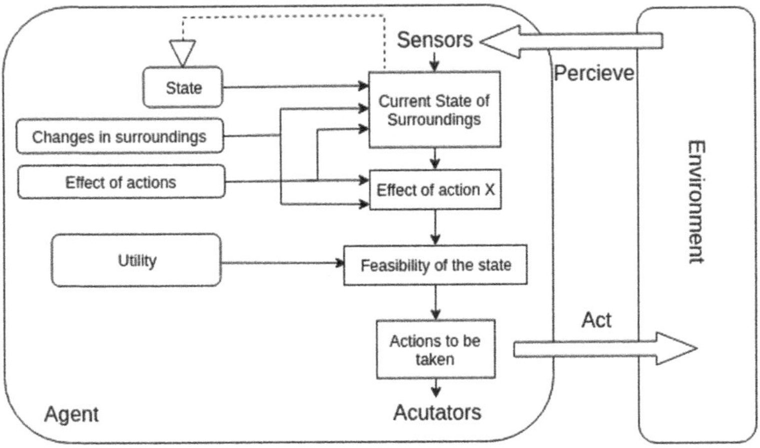

Fig. 9 Utility-based agent

A well-known example of Utility-Based Agent is the Mars Lander. It is an agent-based system sent to Mars to collect samples of rock. If an obstacle is found on its path, implementing Mars Lander as a goal-based agent would result in the agent executing an action in order to achieve the goal without considering if the action is feasible or not whereas implementing Mars Lander as a Utility-based Agent would result in the best path chosen from the utility function (cost or safety).

This reinforces the fact that a utility-based agent delivers many benefits in the aspects of learning and flexibility. Moreover, in certain cases, when goals are insufficient, a utility-based agent will still be able to make rational decisions. Utility functions help in providing a tradeoff between conflicting goals such as the choice between cost and safety of systems. Utility-based agents can act as the best measure to measure the potential of success against the prominence of a goal, in situations where there might be several goals with uncertainty in accomplishment.

4 Multi-Agents in Reinforcement Learning (MARL)

4.1 Learning Agents

A learning agent is capable of learning from its experiences. Initially, it begins with a certain amount of basic knowledge and then becomes capable to act and adapt independently through learning, to enhance its own performance. Unlike simplex agents that act based on information that is provided to them, learning agents are capable of performing tasks, evaluating its own performance and also can search for new ways to become better at those tasks in the future.

A nature-inspired example of learning agents is infants. Infants tend to mimic the behavior of learning agents when they start learning about new things as they grow up. They improve their knowledge base as and when they realize their mistakes in performing various tasks, which helps them understand what actions are to be taken for specific information perceived from its surroundings in the future to avoid repetition of the same mistake.

Learning agents possess the capability to grow from the knowledge perceived from its environment. Their ability to learn allows the agent with no prior knowledge about the environment to become proficient in adapting and understanding its environment. This skill of learning and improving its knowledge base is what makes it to be considered as a "being" with intelligence which is also called "Reinforcement learning". It mainly consists of four components (shown in Fig. 10) which help it in the process of learning, which include *learning element, critic, performance element* and *problem generator*. The *learning element* is accountable for generating improvements, which gives the agent the ability to develop its knowledge base. The *performance element* decides what external action the agent should take based on what it perceives and later shifts to new actions according to feedbacks and suggestions for improvement. The *critic* identifies the outcome of an action, whether it is a reward or a punishment and provides the feedback. The learning element uses this feedback to evaluate the performance of the agent and determines the modification of the performance element, to do better in the future. It has to understand the difference between the performance standards and feedbacks from critics and enhance its standards which will help the agent become competent. The *problem generator* is accounted for experimenting new ways and actions that an agent should try and

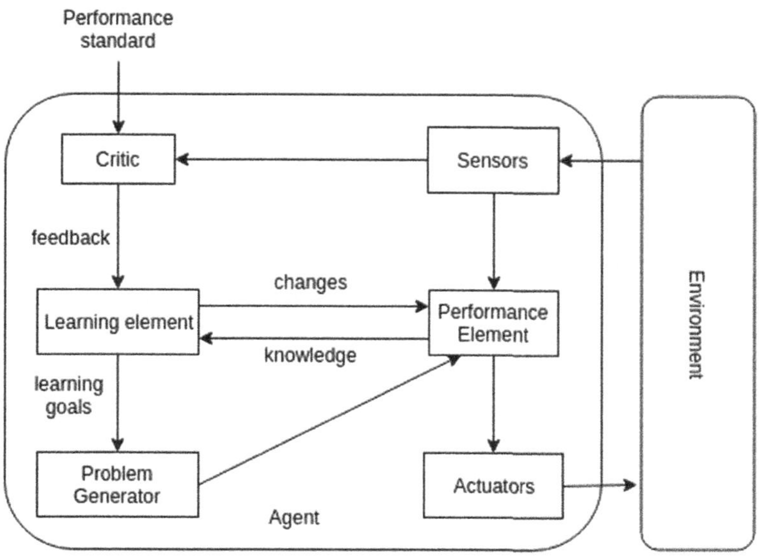

Fig. 10 Learning agent

explore to learn new things about its environment and maybe find an optimal and better set of actions on the long run apart from the given set of standard actions, this enables the agent to learn new informative experiences.

4.2 *Combination of Learning Agents and Reinforcement Learning—Origin of MARL*

The idea of learning in agents lead to the growth of discipline in deep learning which used the concept of reinforcement learning and agent-based systems that could enable the agents to learn and cooperate with its environment and coexist as an intelligent individual. This learning agent has the ability to learn from its mistakes and exhibits complex interactions with its environment based on what it perceives and learns from its experience. This concept was further developed by including multi-agent systems rather than single-agents, that emerged as Multi-Agent Reinforcement Learning (MARL) [16].

A multi-agent system is a set of autonomous, cooperating entities having a shared environment, which they perceive with sensors and based on which they decide their actions and act using their actuators. A reinforcement learning agent learns by coexisting and cooperating with its changing environment. The agent perceives the state of the environment and takes an action, which results in the transition of the environment into a new state. Quality of each transition is decided and evaluated by a scalar reward, and the objective of the agent is to maximize the total rewards throughout the course of interaction. This section throws light on the concept of Agents with Reinforcement Learning and the discipline of Multi-Agent Reinforcement Learning, its benefits, and challenges.

Multi-Agent Reinforcement Learning (MARL) is a field of *Deep Learning* that concentrates on multi-agents and their interactions with their environment and how they can learn from these dynamic interactions. In the context of Single-Agent reinforcement learning the state of the environment changes exclusively because of the actions by a single-agent. These agents after performing a particular action gets the state update of the environment and also gets a scalar reward as a result of its action which tells whether the action is feasible and if it is to be corrected or not. This lets the agent learn from its mistakes and correct its errors and also improve its knowledge as shown in Fig. 11.

In case of MARL referred in Fig. 12, the actions of many agents are reflected on the environment which considers all these actions for transition into a new state. long with this each agent gets back a specific reward and response individually. A MARL environment is expressed as a tuple $\{(\mathtt{P}_1, \mathtt{A}_1), (\mathtt{P}_2, \mathtt{A}_2), \ldots, (\mathtt{P}_n, \mathtt{A}_n)\}$ where P_n represents any given agent and A_n represents its action, then the new state that the environment transitions into is given by the set of combined actions of all the agents represented as $A_1 \times A_2 \times, \ldots, A_\mathrm{n}$.

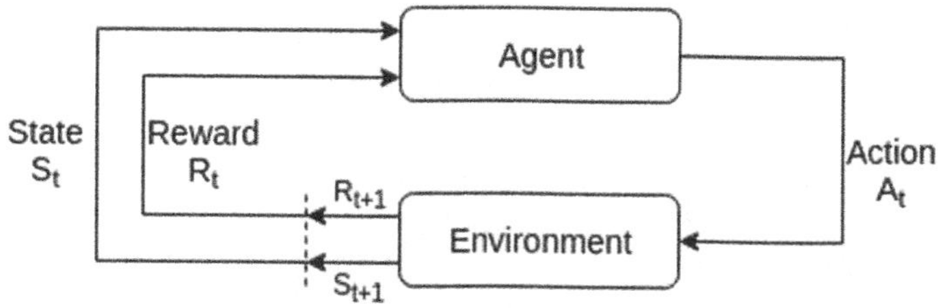

Fig. 11 Single-agent reinforcement learning

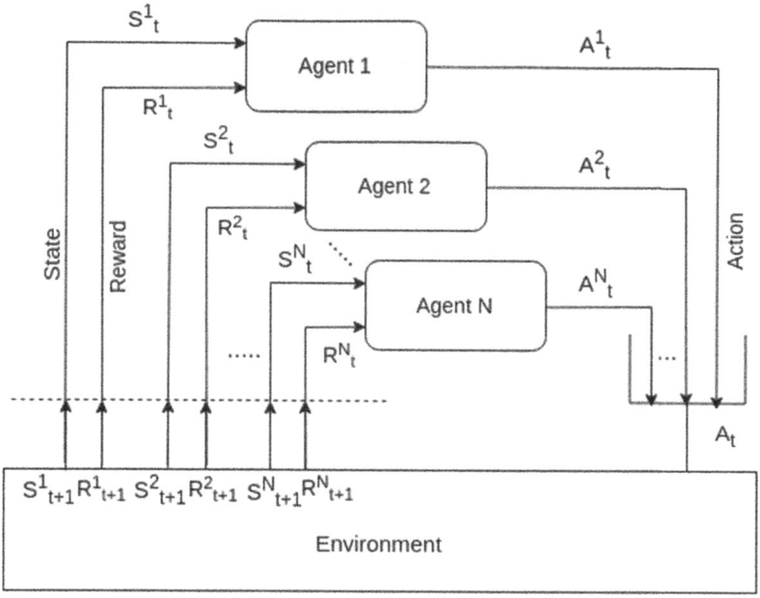

Fig. 12 Multi-agent reinforcement learning

4.3 Demonstrating Learning Agents and Reinforcement Learning Using Netlogo

This model demonstrates an agent with a goal to reach final destination from its source through a maze surrounded by water and ponds within the maze. It provides a practical case study for topics explained in Sect. 4.2. In Fig. 13 the agent begins from its source (black square at bottom-left corner) and it has to reach its destination (pink box at top-right corner). The agent makes multiple trials in order to reach the destination. In this process, it falls into the water and ponds several times and is rewarded with negative scores based on path length or steps that it took to end up in water. Sooner the agent falls into water, greater will be the negativity of score. While on the other hand, when the agent reaches the destination the reward is positive.

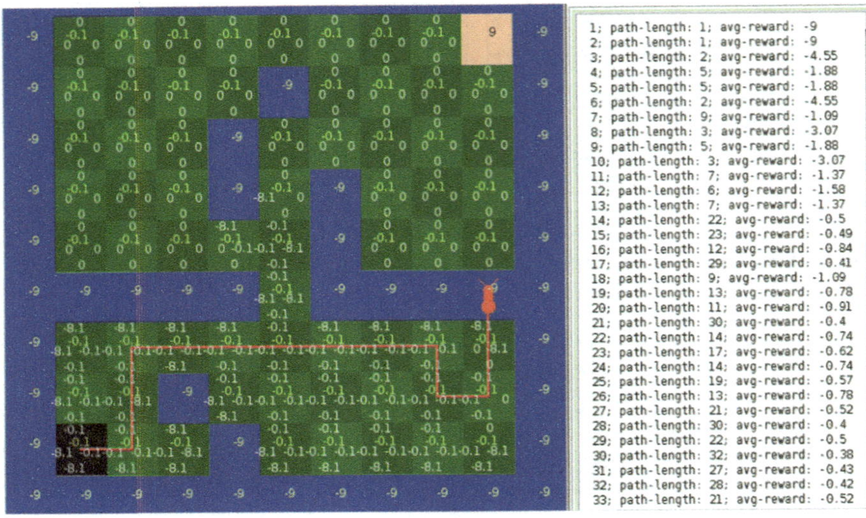

Fig. 13 Agent getting negative reward (in water)

Through these sets of rewards and punishments, the agent learns its path from source to destination and ultimately tries to find the best possible route with an aim to maximize the positive reward resulting in optimal path length [17]. In Fig. 13, it is observed that the agent falls into the water in a path length of 1, meaning it started off in the wrong direction and headed straight toward water, so it is rewarded with a negative score of −9. After several trials, the agent reaches the destination with the best possible route and gains maximum reward score, as shown in Fig. 14 where the reward is 0.22 and path length as 28.

4.4 MARL and Game Theory

The complexity of MARL scenarios is directly dependent on the number of agents present in the environment and also on the kind of behavior that these agents exhibit among each other in the environment. The agents in MARL model can coexist, compete or show neutral behavior. In order to deal with these complexities, MARL techniques adopt some ideas from game theory which helps in modeling environments with multiple agents. Generally, most of MARL models can be demonstrated and expressed using any of these game models [16, 18]:

- **Static Games**: It is a game in which every single player has to choose a strategy simultaneously and they should be unaware of the strategies that are being chosen by other players. Although the strategies might be decided at different moments, the game is still considered to be simultaneous since each player has no idea about

Fig. 14 Agent getting positive reward (at destination)

the strategies of others, hence it is considered as the strategies are being decided simultaneously.

- **Stage Games**: In a static game, a condition arises where the rules of the game depend on the specific stage and condition, at this point the game is known as a stage game.
- **Repeated Games**: When a similar stage game is played numerous times then the game is considered to be a repeated game. Unlike any game that is played once, a repeated game provides scope for a strategy to be dependent on previous moves or actions.

4.5 Challenges of MARL

MARL models offer tangible benefits to Deep Learning tasks given that they are the closet representations of many cognitive activities in the real world. However, there are plenty of challenges to consider when implementing these types of models. Without trying to provide an exhaustive list, there are three challenges that should be top of mind of any data scientists when considering implementing MARL models [16]:

- **The Curse of Dimensionality**:

The most common challenge in the area of deep learning is the problem of dimensions, as the growth in dimensions leads to an exponential increase in their computational complexity. The curse of dimensionality occurs due to the exponential growth of individual state-action space among numerous combinations of state and action variables possible for an agent. When a particular action results in a particular state change this gets added as a new set of dimensions for an agent. As discussed before, the complexity of MARL increases exponentially with the count of agents involved in the model or system, because each agent contributes its own set of variables to the combined state-action space. Hence it happens to be a much bigger challenge in MARL when compared to single-agent RL.

- **Coordination and Correlation**:

Coordination among agents is necessary since the effect of any agent's action on the environment relies on the actions decided by the other agents as well, hence it becomes essential that the agent's choices of actions have to be mutually consistent so that they can achieve their desired effect. It is also important for these agents to have a good correlation in order to understand the behavior and changes of other agents to maintain stability in the environment. So coordinating the training across a large number of agents is another big challenge in MARL models.

- **Ambiguous scenarios**:

MARL models are highly susceptible to ambiguous scenarios such as when two agents attain the exact same condition in the environment. To overcome this, the strategy of each agent has to consider the actions performed by other agents.

4.6 Benefits of MARL

Reinforcement learning agents which perform similar tasks can help each other to learn faster and perform better by sharing their experience. Agents can either share information among each other, can follow other skilled agents or a skilled agent can instruct other agents on how to perform better. Along with better performance, MARL also performs faster, as the agents are involved in parallel computation and work in a decentralized structure in which the task is shared among many agents in the model [16]. Due to this, in any condition of the system where one or more agents may fail, the remaining agents can take over some of their tasks and this makes multi-agent systems more reliable and robust by its characteristics. Multi-agent systems are also scalable as new agents can be easily setup and installed into their systems.

These benefits can be exploited if MARL algorithms are fine-tuned with additional preconditions and prerequisites that each algorithm needs to perform better for a specific task. Certain parameters and steps have to be suitably adjusted and followed which help in providing better results from agent systems. This area of fine-tuning and deciding the conditions in itself is an active field of study. The next section gives an insight into various MARL algorithms and certain optimization techniques that can improve multi-agent systems.

5 Equilibrium Algorithms for Multi-Agent Reinforcement Learning (MARL)

The previous sections have discussed the various types of Multi-Agent Systems, the environment for different types of systems, how each agent in the system interacts with the other agents, and how agents learn from these interactions. But even knowing how the agents interact is not very helpful for practical applications as the environment is ever-changing due to the dynamic nature of the agents. At this moment of time, a system in a state of equilibria would help us greatly. So, in this section, a few algorithms that can help achieve a state of equilibrium in the system have been discussed.

"*Equilibrium*" means a state of rest or no movement in the system. Every environment or system will have its own definitions of equilibrium, so while it is impossible to form a specific equation to achieve a state of equilibrium in all the systems, we will discuss the steps that can be followed will help us in achieving equilibrium. The next subsection introduces the reader to *Q-Learning*, the core concept used in Equilibrium Algorithms for MAS, its mathematical modeling, implementation using NetLogo as well as variants of Q-Learning such as Min-Max and Nash. At the very root of Q-Learning is an equation Q, which will be formulated based on the given constraints and the problem statement of the environment or of the system.

5.1 Q-Learning

In general, the following scenario occurs:

- An environment that consists of a start state and various other states along with multiple agents.
- In each state, an agent can choose from a variety of different actions.
- Each action taken by an agent has its own unique reward.
- After this action is performed, the agent can be in a new state or the same state.

Given the above scenario, let us have a look at Q-learning which is the underlying concept used in all other equilibrium algorithms. Q-learning is an algorithm that uses reinforcement learning and does not depend on the type of multi-agent system model being used. Due to this, it is often referred to as a "model-free" algorithm [19]. The main aim of Q-learning is to come up with a policy or a set of rules that govern what an agent should do under various circumstances when given more than one option. This also means for every change in the environment, a policy (or rule) is already present which tells the agent what action to take. Hence, this approach aims to handle the stochastic nature of the Multi-agent models without having any need for future alterations to the policy. The policy is meant to optimize (either maximize or minimize) the total reward of the system starting from the starting state or any other state of the system. Hence Q-learning returns the reward in the reinforcement so that the agents can learn the best they can.

5.1.1 Equations Involved in Q-Learning

Let "Q" be a function which takes as input the state denoted by the letter "s" and the action denoted by the letter *a*. The function returns a value which gives a quantitative measure of how good that particular action is since this is a learning algorithm, *Q* gets updated each time. So the formula would be:

$$Q(s,a) \leftarrow (1 - \alpha)Q(s,a) + \alpha(r + \gamma \max Q(s,a)) \tag{1}$$

where denotes the discount rate which has a value in the range (0, 1).

denotes the learning rate which has a value in the range (0, 1). (a good value is around 0.1–0.3).

Using the above formula to get the optimal value of the function *Q*,

$$V(s) \leftarrow \max Q(s,a) \tag{2}$$

The outcome of Q-Learning is the actions chosen by the agent. It can be one of two ways:

- Exploration: The actions are chosen randomly.
- Exploitation: The actions are chosen according to max $Q(s, a)$.

5.1.2 Demonstrating Q-Learning Using Netlogo

The model consists of a self-learning agent that has to reach the dark grey shaded box which results in a reward of a hundred points, avoiding going through the light grey shaded boxes which either punish the agent by deducting a hundred points from its scores or provides no reward. `[a, b, c, d]` mentioned for each cell indicates the score received by the agent on moving `[West, North, East, South]` from the cell it currently is in. Initially, in Fig. 15, each box (except for the dark grey and light grey boxes) have `a, b, c, d` as 0 as the agent does not know which direction to move from once it is in any of the boxes. In Fig. 16, after 200 iterations, it is noticed that the values which were previously zero now have either a value greater than 0 or a value less than 0. The more positive the value is, the more likely it helps to achieve the goal. After iteration 400 in Fig. 17, it is seen that the direct (or more optimal path) for the agent has started revealing itself as those boxes and the directions the agent is supposed to move in now has higher positive values. At the end of iteration 600 in Fig. 18, the path the agent needs to take is pretty obvious, thus the agent has now learned from its given environment which path it's supposed to take to reach its final goal. The number of iterations taken to reach this greatly

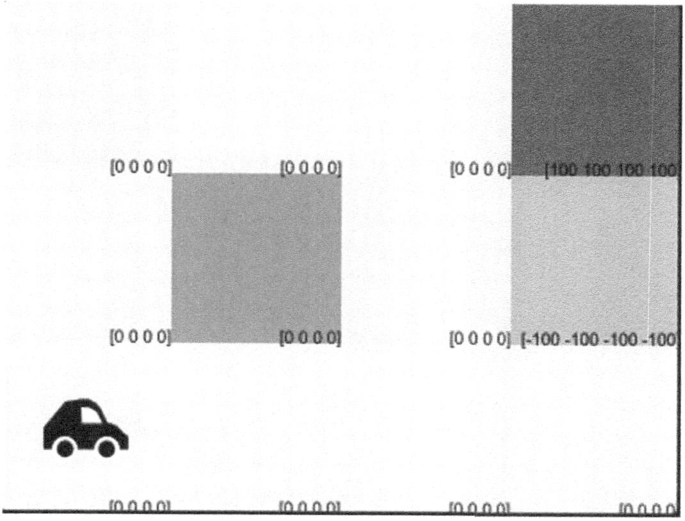

Fig. 15 Iteration 0

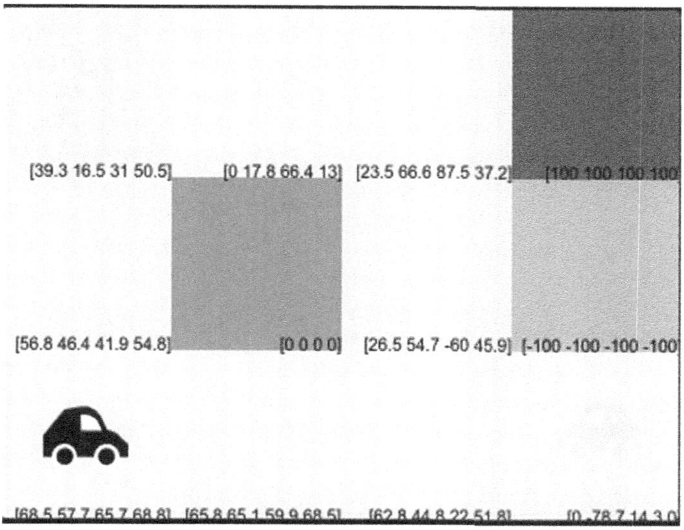

Fig. 16 Iteration 200

depends on the Q-function and the learning rate chosen for the agent. These values are decided at the beginning (before the second iteration) and is applied to Eq. 1 so that the agent learns from the previous experience and improves its decision-making ability. To run the model, parameters γ and α for the Eq. 1 were set to 0.9 and 0.5, respectively [20].

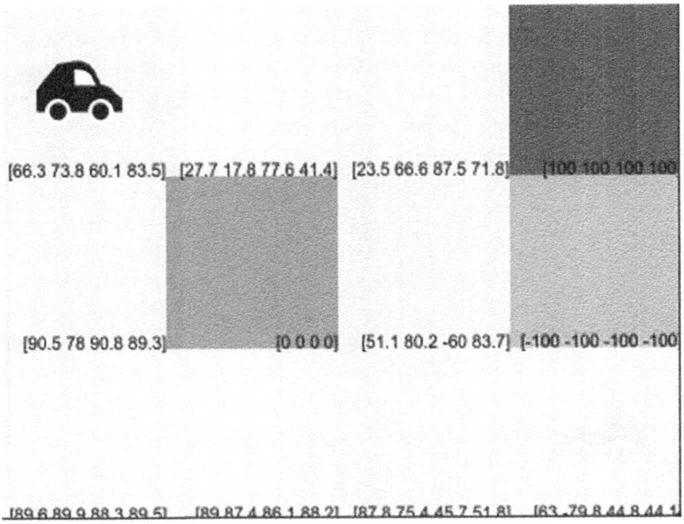

Fig. 17 Iteration 400

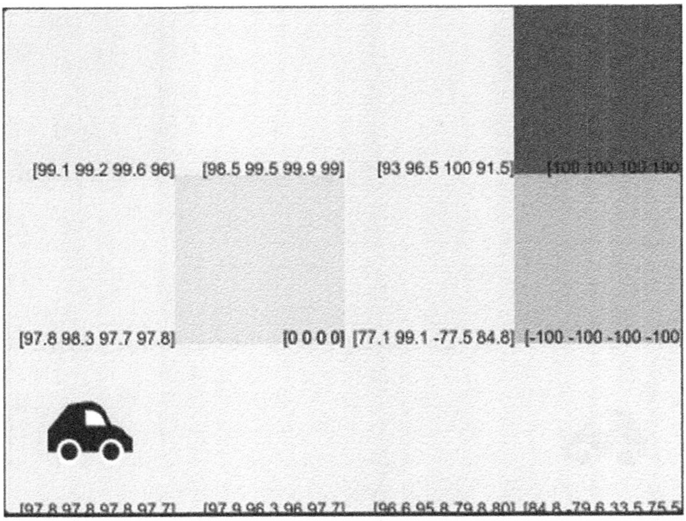

Fig. 18 Iteration 600

5.2 Minimax Q-Learning—A Popular Q-Learning Variant

Consider a two-person game in which each player is an agent. Here each player can take one action out of the two available actions: A or B. Applying Q-learning naively would not work well for us in this situation as [21]:

- If the opponent takes a complex strategy, would be very hard to quantify the results of the function Q.
- There is no guarantee of convergence of the Q-learning algorithm at some value, due to which we could be stuck in an infinite loop.

To solve the above problems, we apply the minimax Q-learning approach. As the name suggests, minimax (or maximin) means choosing the minimum from a range of values where each value in this range is the maximum of some particular function. Hence, we call this minimax (choosing the minimum from the maximum). Maximin is the vice versa of the above.

In Minimax Q-learning, in addition to the state s and action a (as seen as in Q-learning), we also consider what action the other agent (in this case the opponent) can take and denote it with the letter o [21]. The updated formula would now be

$$Q(s,a,o) \leftarrow (1 - \alpha)Q(s,a,o) + \alpha\left(r + \gamma V(s')\right) \quad (3)$$

where Q, s, a, o, , and have the same meanings as in Eq. 1.

We use the above formula until we get the optimal value of the function Q

$$V(s) \leftarrow \mathrm{maxmin} \sum Q(s,a,o)\pi_s(a) \quad (4)$$

where Π denotes the probability of choosing some action when following some strategy π_s

Here, the actions are chosen by

- At the beginning the set of π_s is used to choose a random action for each state.
- Before each step, exploration (random action) and exploitation (maxQ(s,a,o)) are used.
- After each step, the values are updated to π_s so that the MaxiMin strategy can be applied based on the results obtained using Eq. 3.

Minimax Q-learning performs better than naive Q-learning not only in efficiency but also by assuring a convergence to various values. The only drawback here is the rate of convergence which might be very long for some cases [21].

5.3 Nash Q-Learning

This Q-learning algorithm tries to achieve Nash equilibrium which is "A stable state of a system involving the interaction of different participants, in which no participant can gain by a unilateral change of strategy if the strategies of the others remain unchanged." [22].

Agent i's Nash Q-function is written over all of the states like the sum of the Agent i's current reward and its future rewards when all of the agents follow a joint Nash equilibrium strategy. Mathematically, it is denoted as →

$$Q_i * \left(s, a^1, \ldots, a^n\right) = r^i\left(s, a^1, \ldots, a^n\right) + \beta \sum_{s0 \in S} p\left(s_0|s, a^1, \ldots, a^n\right) v^i\left(s_0, \pi_*^1, \ldots, \pi_*^n\right) \quad (5)$$

where π_i denotes the joint Nash equilibrium strategy the agent has chosen,

r_i (s, a^i) is the agent i's one time reward in a state s and under the joint actions a_1 up to a_n

v_i (s_0, π^i) is the agent i's total discounted reward which is calculated over all the states starting from state s_0 assuming that at any given time, each agent follows the equilibrium strategies [22].

The optimal Q-value is obtained by using Eq. 5 recursively. Apart from Minimax and Nash Q-learning algorithms, there are other algorithms like relief based, correlated equilibrium, Pareto-optimality and so on which are beyond the scope of this book.

5.4 Policy Hill-Climbing

Another algorithm that uses Q-learning to play mixed strategies in multi-agent stochastic systems is policy hill-climbing (PHC) [23]. The principal of this algorithm can be understood taking a Hill as an example. A hill to a great extent can be imagined as the graph of a n-degree polynomial equation. Given any random point on this graph (or hill), we check the corresponding points in the neighborhood (that is, point with slightly smaller and slightly larger values) and check the corresponding Q value. We then move toward the higher Q value and after a particular number of iterations, we reach the global maxima or minima hence achieving a state of equilibrium. Hence this algorithm is called Policy Hill-Climbing.

This method performs hill-climbing in the space of mixed policies. Here, Q values are maintained just as we see in the naive Q-learning but here, along with that, PCH also keeps track of the current mixed policy. This policy is improved by increasing the probability that it selects the highest valued action according to a learning rate alpha. As the value of alpha approaches 1, the algorithm starts behaving as our traditional Q-learning, and when = 1, it is completely the same as Q-learning. This method just like Q-learning is not only rational but will also converge to an optimal policy as long as the other agents are following only one particular strategy instead of a complex strategy [23].

6 Optimization of Multi-Agent Systems

Optimization in Multi-Agent System (MAS) has given us a large collection of techniques to solve many of the problems that exist in the various application domains which involve multi-agent systems. The solving of these problems requires the active participation of the agents in a Multi-Agent System. Research on Multi-Agent System optimization has quickly become a very technical and specialized field [2]. So in this section, let's have a look at some of the problems faced by MAS developers and some methods which help optimize the Multi-Agent System.

6.1 Issues that MAS Developers Deal With

A multi-agent system developer faces the following issues →

- Maximizing the performance of all agents.
- Making sure all the agents are used to the fullest, that is, they try to ensure that no agent is left idle at any moment of time.
- Minimizing time and space complexities.
- Minimizing communication cost between the agents(when groups or coalitions are considered).
- Ensuring that all the agents are working in the same direction and no agent is working against the requirements.
- Allocating the optimal task to each agent to maximize the overall performance.

6.2 Distributed Constraint Optimization (DCOP)

A DCOP is defined as a state where one or more agents must be assigned values for a set of requirements in such a way that the total cost of the requirements over a finite number of given constraints is minimized [24]. Thus, optimizing the available resources and in turn, the multi-agent system. The constraints are for some parameters which are part of a domain and need to be assigned to the same values by all the agents. There are multiple algorithms designed to solve these types of problems.

Generally, a DOCP is described as a tuple (A, X, D, f, ,) where

- A is the set that contains all the Agents.
- X is a set of all the variables in the system $\{x_1, x_2, \ldots, x_n\}$.
- D is a set of domains which the problem statement deals with $\{D_1, D_2, \ldots, D_n\}$.
- F is a function that gives every variable (taken as input) some cost (output).
- is a function that maps all the variables to their associated agents, : $V \rightarrow A$.
- is a symbol that adds all of the values of f when a different variable x is assigned.

That is, $\eta(f) = \sum f(s)$, where s is the various combinations of variable and domains $s \in (D_i \times V_i)$.

The entire objective of DCOP is to either maximize or minimize the value of $\eta(f)$ based on the requirements.

Various DCOP algorithms and their memory complexity is shown in Table 4.

Table 4 Various DCOP algorithms and their memory complexity

Algorithm name	Memory complexity
Distributed pseudo tree optimization procedure [34]	Exponential
No-commitment branch and bound [34]	Polynomial
Asynchronous partial overlay [34]	Polynomial
Asynchronous BackTracking [34]	Polynomial

Example: Distributed Graph Coloring
This is a well-known problem in discrete mathematics where the goal of the problem is to divide the available vertices into different sets of vertices with the help of coloring such that no two adjacent vertices possess the same color. Here, each vertices can be considered to have an agent which decides its color. Each agent has a variable whose associated domain is of cardinality $|C|$ (The number of colors). For each of the vertex agent, a variable v_i is created with domain $D_i = C$. [25] A constraint on cost is then created, that is if both the variables have the same value, cost is 1. The objective then becomes to minimize the cost function $\eta(f)$.

6.3 Coalition Formation Algorithms

This section talks about how multiple agents when given multiple tasks divide themselves into various coalitions and how they perform the various tasks given to them. This task allocation to groups of agents is necessary as often is the case that one agent cannot solve the task alone. Also, when multiple agents perform on a single task, this provides better performance as compared to a single-agent performing the same task.

The main goal of the coalition algorithm is [26]

- Allow all of the agents to make different groups and to assign tasks to all of the groups.
- Enable distributed allocation.
- Lower the computational complexity as shown in Fig.19.

Try it yourself: Draw the coalition graph for a 4-agent system (Hint: There are 15 combinations possible). An example for a 3-agent system is shown in Fig. 19.

Agents are always Group-Rational [26], i.e.,→

- Only if the performance of the coalition is greater than the performance of the individual, coalitions are formed. If this is not true, agents don't prefer to form coalitions.
- As a result of whenever agents form a coalition, the system's outcome is always greater than the outcome without the coalition. The new outcome is the sum of all the individual coalitional outcomes.

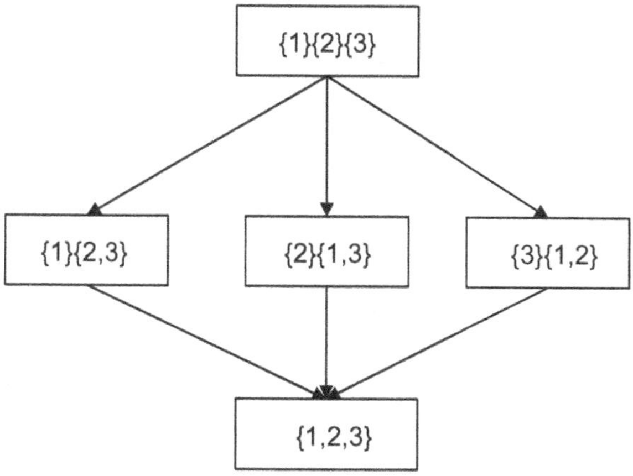

Fig. 19 Coalition graph for a 3-agent system

An environment is called a Super-additive environment if the following rules are satisfied:

→ For disjoint C_1, C_2 in the set, if they join together, then $V_{new} >= V_{c1} + V_{c2}$

The above property thus promotes the agents to form larger coalitions.

While some environments promote larger coalitions, in practical usage small size coalitions are better as:

- Communication and computation times are less expensive.
- They are more economical as for most tasks, small number of agents are enough to accomplish it.
- This heuristic will limit the coalition size to be an integer k.
- Number of coalitions will be $O(n^k)$

Formally we can define a coalition system as [26] →

- A set of *n* agents, $A = \{ a_1, a_2, \ldots, a_n \}$
- Each a_i has a vector of real positive capabilities $B_i = < b_1^i, \ldots, b_r^i >$
- A set of independent tasks, $T = \{ t_1, t_2, \ldots, t_m \}$. Each task t_i is associated with a vector of real positive capabilities.
- A coalition *C* has a vector of capabilities B_C which is the vector obtained when we add each of the individual capabilities of the members in the coalition.

7 Applications of Multi-agents—Case Studies

In recent decades, MAS has become a powerful tool for engineering applications and in the realm of Decision science. As a computational paradigm, MAS is the elucidation for distributed control [27]. Few widely used application of MAS is discussed briefly in this section.

7.1 Multi-agents to Build an Optimal Supply Chain Management

Extensive usage of Multi-agents lies in Supply Chain Management (SCM) [27]. A supply chain is a network of suppliers, manufacturing plants, warehouses, and distribution channels that performs functionalities including procurement of raw materials, transformation of these materials into finished products and the distribution of these products to customers.

The Multi-agent-based architecture strives to build an optimal supply chain management system since Multi-Agent systems enable increased autonomy of each participant in the supply chain. Each agent is expected to analyze and arrive at an optimal solution, striving for individual goals while aiming to satisfy both the local and the global constraints of the system. Therefore, one or several agents can be used to represent each participant in the supply chain mentioned in Table 5. Moreover, the agent paradigm is a natural metaphor for network organizations, since companies' end goal is always in maximizing their own profit. Therefore the participants have the same characteristics as agents mentioned in Sect. 2.1 and hence can be implemented using agents that optimize the supply chain [27].

Q-learning discussed in Sect. 5.1 can be applied for the MARL agents to implement SCM with the objective of each agent being minimization of the cost of the inventory of the overall supply chain [28]. Q-Learning algorithm applied for SCM agents is given here:

- *Initialize estimated values **Q(s, a)** for each state "s" with its associated action "a"*
- *Do until end condition is met:*

  a. ***s(t)*** ← *current state of the agent at time t in the environment*
  b. ***a(t)*** ← *current action selected according to action-selecting policy and executing it*
  c. ***s(t + 1)′*** ← *next state resulted on executing the action*
  d. ***r(s(t), a(t))*** ← *reward function given by* r_t `=∑j∑i [ ICij(t) + BCij(t) ]` *at time t, i.e., a function for computing the reward based on holding inventory and backorder costs at time t.*

Table 5 The key participants in a supply chain and problems faced by them

Participant	Role	Problem to be optimized
Suppliers	Provider of goods and services	Procurement of components
Manufacturers	Engaged in the original production and assembly of products, equipment or services	Production of finished goods on time
Distributors	Entities that sell the product to the customers and collect payments	Finding customers and delivering the finished goods

e. *Update value* $Q(s(t), a(t))$ *according to* $Q_t(st, at) = (1 - \alpha)Qt - 1(st, at) + \alpha[rt + \beta maxbt\ Qt(st + 1', bt)]$ *where* $Q(t)$, st, β, at, α, rt *representing estimated value, state, discounted factor, an action, learning rate and reward, respectively at time t.*
f. *Update state* **s** *to new state* **s**′.

The reward function is derived in order to achieve the goal of the SCM system (minimizing the inventory cost) [28]. *Q*-values converge to the estimated optimal values in each state on repeating the above process for multiple iterations. On completing the process of learning, the best policy of each state can then be derived by the action-selecting rule.

Another approach could be to implement each agent in the system by choosing any of the agent types explained in Sects. 3 and 4.1, which provides the designer flexibility to implement heterogeneous strategies and diverse types of agents to build a system. The optimal combination of these different agents working together as a system can be decided further upon testing. The agents can also collaborate with each other by sharing one's results using a distributed knowledge base in order to minimize the cost of the SCM system [29].

7.2 Optimization Technique Using Ant Colony Based Multi-agents for Traveling Salesman Problem

Traveling salesman problem (TSP) is an important NP-hard optimization problem. TSP is the enigma of a salesman who commences his journey from his hometown seeking the shortest tour route through the set of consumer cities in his itinerary and then concludes his journey by returning to his hometown, paying a visit to each consumer city precisely once. Formally, the goal of the TSP is to find the closed path in a graph *G* visiting each of the n nodes of G exactly once aka Hamilton circuit. Due to its computational intractability, it has allured several heuristic advances to produce satisfactory, if not optimal solutions.

Ant Colony Optimization (ACO) is a nature-inspired optimization [6] algorithm that parodies the behavior of ant foraging [30, 31] based on important insights observed from the ants' behavior

- Ants begin to explore for food in a random manner around their nests.
- As soon as they find their food, they travel back home dropping continuously a chemical substance produced by them called pheromone that guides the rest of the ants to reach the locus of the food.
- With growing number of ants picking the trail with high pheromone intensity, they reach the food source quickly.
- Most of the transmission of information among the ants, or between ants and the environment, is based on the exploitation of pheromone rates produced by the

Rule 1 - If destination is in the ant's detectable range, then he must directly move towards it, otherwise he moves to the place with highest intensity of pheromone. If the intensity of pheromone in his neighbour cities are all the same value, then he can randomly choose a direction to move. If the pheromone values within his neighbour cities are all zero, then he will keep his advance direction.

Fig. 20 Rule 1

Rule 2 - If an ant has covered all the cities, then he returns to his nest at once as he continually deposits pheromone on the path which evaporate with time.

Fig. 21 Rule 2

ants. Using this the ant colony is competent to find the optimal path in a relatively shorter duration of time [32].

7.2.1 Applying the Procedure of ACO Based Multi-agents to Solve TSP

Each ant is an independent agent that shares the common objective of finding the shortest route through the cities. These agents have no global knowledge of the whereabouts of the optimal solutions but benefit from the advantage of communicating through pheromones that help the agents to share the information. These agents need to accomplish the task by following certain rules mentioned in Figs. 20 and 21 which are the "`If state ->Then action`" rules that are needed to be designed in simplex-reflex agents discussed in Sect. 3.1. Using these sets of rules, the agents can find the optimized path to cover all the cities over the course of iterations [32]. The MAS based on ant colonies exhibits the proficiency of habilitating to the changing conditions and unknown environments by perceiving the information of pheromone on the paths from its environment and taking necessary actions according to simple rules.

Figure 22 shows the procedure of applying ACO based MAS. After initializing the number of ants and the intensity of pheromone equally on all paths, and designing the simple local rules, the ACO algorithm iterates until all the cities are covered and the ant is back home, in which all of the ants' tours are constructed after matching with the rules and finally the pheromone trails are updated.

7.2.2 Demonstrating ACO Based Multi-agent Model with Netlogo

According to the predefined rules, each agent (i.e., ant) must make a choice at each node as to which is the next node to be traveled, structuring a tour route in the graph

Fig. 22 Procedure of applying ACO based multi-agents for finding shortest route in TSP

```
function TSP returns shortest-optimal-path
  Initialize pheromone trails
  Set the number of ants
  Initialise Rules
  while (termination condition not met) do
    ApplyRuleMatch
    ConstructAntsSolution
    UpdatePheromones
   end
  end

return shortest-optimal-path
```

covering all the nodes. At any iteration, the probability of an agent choosing the next node is determined by the intensity of pheromone on the path and the distance associated with each edge. The model implemented consisted of 20 agents and these agents had to find the cost-effective path for TSP covering 20 cities (nodes). In Fig. 23, the pheromones are equally assigned to each path. Hence the agents move in a random fashion. In Fig. 24, it is noticed that after five iterations the agents are making decisions based on pheromone rate on the path and in Fig. 25 the optimal path solving the TSP found by the best agent is depicted [33].

Figure 26 shows a plot for performance analysis which depicts the cost of the shortest route found in the TSP over the course of iterations.

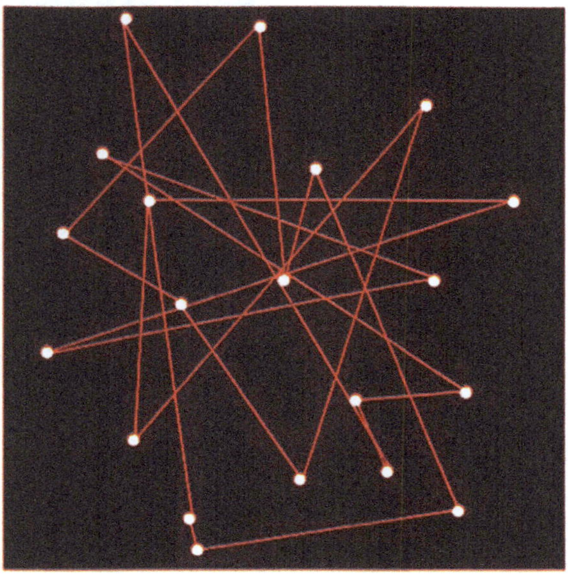

Fig. 23 Iteration 0

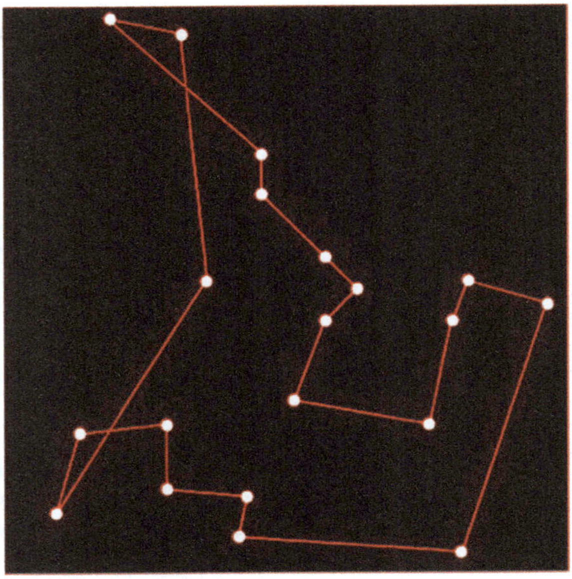

Fig. 24 Iteration 5

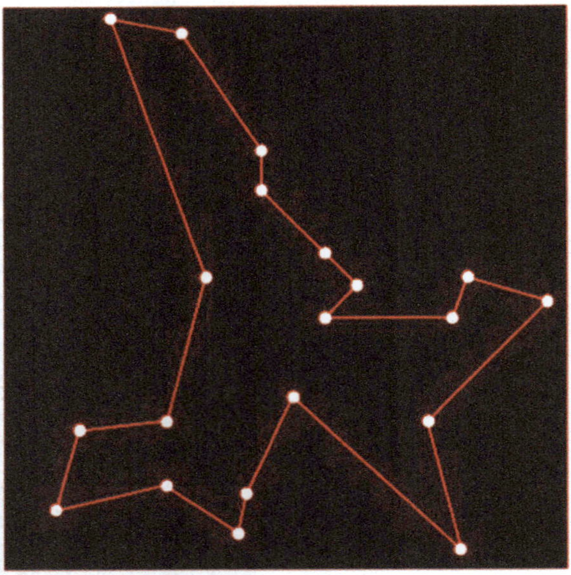

Fig. 25 Iteration 87

8 Conclusion and Research Trends

Research into MAS using reinforcement learning has been picking up in the last decade and now that a strong foundation has been laid in this area, more and more

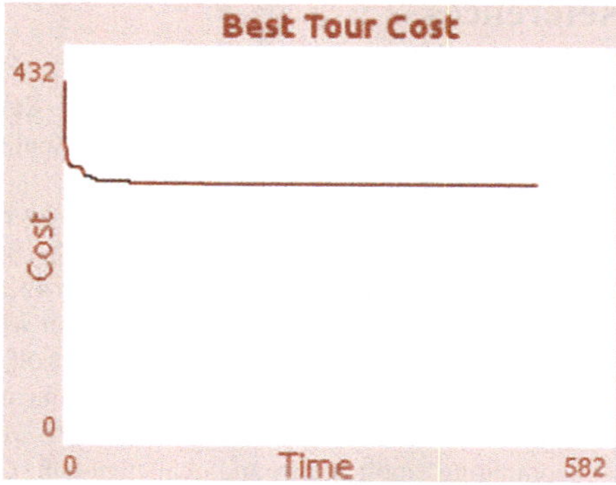

Fig. 26 Best tour cost

new and useful applications are being devised. This growth can be attributed to the following reasons:

- MAS denotes perfect examples of real-world situations and ecosystems where multiple agents interact with each other and their surroundings to accomplish a particular task.
- Any living being learns from its mistakes, due to this simple correlation between a living organism and a reinforcement learning model, the applications of MAS are infinite.
- Multiple softwares are being developed purely for research purposes on MAS and how they behave under reinforcement learning. For example, Netlogo.
- Famous languages like Python (Spade, Pade) and Java(Jade) have also created new frameworks that are dynamically being updated to help in understanding MAS better.
- Various conferences like OptiMAS which are conducted annually across the globe help share knowledge on a global level and spread awareness about the new software and algorithms that are made.

Research into MAS and its various applications although being an infinitely vast topic will continue to evolve over time and as our understanding of these interactions increases, we will see more applications in the real world.

To conclude, the methods and models introduced in this chapter aim to give the reader an in-depth insight into the topics involved in MAS. While the number of algorithms and optimization techniques are huge, the principle underlying all of them is the same and has been discussed. At the end of this chapter, it is hoped that the reader has a clear understanding of how different agents interact in different environments and what type of agents to choose under various circumstances while optimizing the agent's performance.

References

1. Xavier overview. http://www.cs.cmu.edu/~Xavier/overview.html
2. COG project overview http://www.ai.mit.edu/projects/humanoid-robotics-group/cog/overview.html
3. Sony Aibo|the history of robotic dog http://www.sony-aibo.com/
4. Zeghida D, Meslati D, Bounour N (2018) Bio-IR-M: a multi-paradigm modelling for bio-inspired multi-agent systems. Informatica 42. https://doi.org/10.31449/inf.v42i3.1516
5. Sneha V., Shrinidhi KR, Sunitha RS, Nair MK (2019) Collaborative filtering based recommender system using regression and grey wolf optimization algorithm for sparse data. In: 2019 IEEE international conference on communication and electronics systems (ICCES) (in press)
6. Nayak J, Naik B, Jena AK, Barik RK, Das H (2018) Nature inspired optimizations in cloud computing: applications and challenges. In: Cloud computing for optimization: foundations, applications, and challenges. Springer, Cham, pp 1–26
7. Das H, Jena AK, Nayak J, Naik B, Behera HS (2015) A novel PSO based back propagation learning-MLP (PSO-BP-MLP) for classification. In: Computational intelligence in data mining, vol 2. Springer, New Delhi, pp 461–471
8. Dey N, Das H, Naik B, Behera HS (eds) (2019) Big data analytics for intelligent healthcare management. Academic Press
9. Sahoo AK, Mallik S, Pradhan C, Mishra BSP, Barik RK, Das H (2019) Intelligence-based health recommendation system using big data analytics. In: Big data analytics for intelligent healthcare management. Academic Press, pp 227–246
10. Xie J, Liu C-C (2017) Multi-agent systems and their applications. J Int Counc Electr Eng 7(1):188–197. https://doi.org/10.1080/22348972.2017.1348890
11. Nagwani NK (2009) Performance measurement analysis for multi-agent systems. In: 2009 international conference on intelligent agent & multi-agent systems, pp 1–4
12. Milewski J, Multi agent reinforcement learning. https://www.cs.ubc.ca/~kevinlb/teaching/cs532l%20-%202013-14/Lectures/rl-pres.pdf
13. Downloading Netlogo (n.d.) https://ccl.northwestern.edu/netlogo/oldversions.shtml
14. Tisue S, Wilensky U (2004) NetLogo: design and implementation of a multi-agent modeling environment. https://ccl.northwestern.edu/papers/2013/netlogo-agent2004c.pdf
15. Junior IC (2013) Data mining with ant colony algorithms. Lect Notes Comput Sci 30–38. https://doi.org/10.1007/978-3-642-39482-9_4
16. Busoniu L, Babuska R, De Schutter B (2010) Multi-agent reinforcement learning: an overview. https://doi.org/10.1007/978-3-642-14435-6_7
17. Roop J (2006) Reinforcement learning in NetLogo. http://ccl.northwestern.edu/netlogo/models/community/ReinforcementLearningMaze
18. Nowe A, Vrancx P, De Hauwere Y-M (2012) Game theory and multi-agent reinforcement learning. https://doi.org/10.1007/978-3-642-27645-3_14
19. Poole D, Mackworth A, Artificial intelligence. https://artint.info/html/ArtInt_265.html
20. Lin L (2014) Q-Learning in MDPs (n.d.) http://modelingcommons.org/browse/one_model/3986#model_tabs_browse_nlw
21. Cerquides J, Farinelli A, Meseguer P, Ramchurn SD, A tutorial on optimisation for multi agent systems. https://academic.oup.com/comjnl/article-abstract/57/6/799/378684?redirectedFrom=fulltex
22. Hu J, Wellman MP (2003) Nash Q-learning for general-sum stochastic games
23. Wilensky U (1997) NetLogo wolf sheep predation model. Center for Connected Learning and Computer-Based Modeling, Northwestern University, Evanston, IL. http://ccl.northwestern.edu/netlogo/models/WolfSheepPredation
24. Picard G, Distributed constraint optimisation. https://www.emse.fr/~picard/cours/mas/lecture-DCOP-2017.pdf
25. Bae J (2019) Distributed graph coloring. Stanford.edu. https://stanford.edu/~rezab/classes/cme323/S16/projects_reports/bae.pdf. Accessed 1 May 2019

26. Tang D (2019) Coalition formation. Cpp.edu. https://www.cpp.edu/~ftang/courses/CS599-DI/notes/Coalition%20Formation.pdf. Accessed 1 May 2019
27. Sardinha JARP, Molinaro MS, Paranhos PM, Cuhna PM, Milidiu RL, Lucena CJP (2005) A multi-agent architecture for a dynamic supply chain management. Monografias em Ciencia da Computacao, No. 36/05
28. Zhao G, Sun R (2010) Application of multi-agent reinforcement learning to supply chain ordering management. In: 2010 sixth international conference on natural computation. https://doi.org/10.1109/icnc.2010.5582551
29. Rudenko D, Borisov A (2006) Agents in supply chain management: an overview. https://ortus.rtu.lv/science/lv/publications/5910/fulltext.pdf. Accessed 30 Apr 2019
30. Dorigo M, Caro GD, Gambardella LM (1999) Ant algorithms for discrete optimization. Artif Life 2:137–172
31. Dorigo M, Maniezzo V, Colorni A (1996) Ant system: optimization by a colony of cooperating agents. IEEE Trans Syst Man Cybern-Part B 1:29–41
32. He J, Min R, Wang Y (2005) Implementation of ant colony algorithm based-on multi-agent system. In: Lu X, Zhao W (eds) 2005 international conference on networking and mobile computing (ICCNMC). Lecture Notes in Computer Science, vol 3619. Springer, Berlin, Heidelberg
33. Roach Christopher (2007) NetLogo ant system model. Computer Vision and Bio-inspired Computing Laboratory, Florida Institute of Technology, Melbourne, FL
34. Fioretto F, Pontelli E, Yeoh W, Distributed constraint optimisation problems and applications: a survey. https://www.researchgate.net/publication/301844403_Distributed_Constraint_Optimization_Problems_and_Applications_A_Survey

Chapter 8
Computer Vision and Machine Learning Approach for Malaria Diagnosis in Thin Blood Smears from Microscopic Blood Images

Golla Madhu

Abstract Malaria is still a serious health problem globally, which is caused by a parasite of genus Plasmodium through Anopheles mosquitoes. An automated and accurate diagnosis is needed for an appropriate intervention to reduce mortality and to prevent anti-malaria resistance. In general, pathologists investigate blood-stained slides through conventional microscopy for diagnosis. However, these approaches are done by a clinical expert which biased, error-prone and moderate. This research addresses the development of robust computer-assisted malaria diagnosis in light microscopic blood images. In addition, microscopic images were obtained through stained slides which consist of illuminations and noise levels. To overcome this situation used computer vision approach i.e., improved *k*-SVD denoising method, fuzzy type II-based segmentation, feature extraction through local and global features, feature selection with Extra Trees Classifier and classify the different stages of malaria have been explored. The proposed classification strategy can be achieved by unifying Extremely Randomized Trees (ERT). The obtained experimental results showed Extremely Randomized Trees classifier accuracy of 98.02% that have been achieved in the process of malaria diagnosis.

Keyword Computer vision · Extra trees classifier · Machine learning · Malaria · Microscopic blood images

1 Introduction

According to World Health Organization (WHO) report, malaria is confirmed as a major global health problem, which is caused by Plasmodium parasites like Plasmodium falciparum, Plasmodium malaria, Plasmodium vivax, and Plasmodium ovale [1, 2]. According to the NVBDCP's report, India is endemic for malaria patients who are affected by Plasmodium falciparum, and vivax are the major infectious disease of the population [3]. In 2017, India reported about 0.842 million positive cases

G. Madhu (✉)
Department of Information Technology, VNR Vignana Jyothi Institute of Engineering and Technology, Hyderabad 90, Telangana, India
e-mail: madhu_g@vnrvjiet.in

J. K. Rout et al. (eds.), *Machine Learning for Intelligent Decision Science*,
Algorithms for Intelligent Systems, https://doi.org/10.1007/978-981-15-3689-2_8

which remain as the maximum infection disease declared by the WHO [4]. Most of the countries follow the 'gold standard' technique for laboratory diagnosis, which is done by microscopy investigation of stained blood smears. In general, we have two types of standard microscopic examination approaches such as thin and thick blood images [5]. Pathologists examine the thin or thick blood images under light microscopy for the malaria diagnosis through color, morphological features, and accuracy depends on the pathologist understating [6, 7]. However, these procedures involve radical errors in terms of bias that leads to less diagnostic accuracy in the diagnosis process [7]. Keeping these limitations, modern digital diagnostic systems have been developed and applied in the diagnosis process of malaria. Further, computer vision and machine learning methods have made a huge amount of contribution towards better diagnosis with the highest accuracy in various medical imaging analyses such as; CT scan, MRI, Microscopy, Ultrasound imaging analysis, and others [5, 6]. In literature, various computer-aided diagnosis methods suggested for malaria diagnosis under light microscopy [6–9]. In malaria diagnosis, texture features and morphological features play a dynamic role in identifying the parasites by unique features. Thus, feature extraction and selection play an important role in the classification of malaria parasite stages. In view of these issues, we concentrate on the development of computer vision and machine learning malaria diagnostic models for discriminating against various stages of malaria parasites.

2 Related Works

Erez and Pearl [10] suggested a standard technique for the preparation of thick blood films in the diagnosis process of malaria. Tek et al. [11] presented an analysis of computer vision and image processing study for automated diagnosis malaria from microscopic blood images of thin blood films. This study addresses various issues of imaging, its interoperability of the diagnosis process and limitations of segmentation of the blood image that have been discussed. Tek et al. [12] proposed an automated malaria parasite detection and testimony of Giemsa stained blood smears. The proposed framework classifies the different infected life stages of malaria with 0.1% false detection. Kumarsamy et al. [13] presented a unique approach for recognition of malaria parasites and its stages classification from digital images. This study detected weak and strong boundary edges in red blood cells (RBCs) using similarity measure and rule-based method used to tie the edge portion that forms closed contours of the RBCs. Finally, it attained a classifier accuracy of 97% for cell segmentation and precision of 86% for detection. Bhowmick et al. [14] discussed a computer vision and geometric morphological feature extraction for erythrocytes detection from scanning electron microscopic (SEM) images. In addition, a marker-controlled watershed technique for the segmentation of erythrocytes of SEM images and multilayer perception (MLP) techniques were used for the classification of these erythrocytes, overall accuracy of MLP is 94.59%. Das et al. [6] addressed a computer-assisted malaria diagnosis for parasite detection and Bayesian learning and SVM are used for the classification of feature vectors. Using a Bayesian approach and SVM provides the highest accuracies is 84% and 83.5% respectively. Linder et al. [15] presented a diagnostic decision support scheme for the detection of malaria parasites utilizing computer vision methods with visualization techniques. In this approach, the linear SVM classification model and the compatible kernel were used for the classification of infected and non-infected malaria parasites in blood images. Somasekar and Reddy [16] presented an edge-based segmentation technique for erythrocytes of infected region detection from microscopic blood images. The infected cell regions are extracted through Fuzzy C-Means (FCM) algorithm and Minimum Perimeter Polygon (MPP) algorithm used for parasite extraction to refine the cell edges in the images. Das et al. [17] proposed image characterization and a classification framework for the malaria diagnosis process and stages detection from microscopic blood smears. This study, erythrocytes overlapping problem handled by marker-controlled watershed methods with a multilayer perceptron network classifier provides the highest classifier accuracy of 96.84% for the stage classification in the diagnosis process. Das et al. [18] presented a PSO based backpropagation neural network for the classification of datasets. Rosado et al. [19] presented an analysis of an automated classification system to assess the malaria parasites detection and its life cycle stages from microscopic blood images. This review most of the approaches are proposed and verified with a part number of images, and more images needed. Bibin et al. [20] suggested an innovative method for malaria detection using deep belief networks (DBN) from blood images. This study, contrastive divergence technique for preprocessing and back-propagation algorithm which measures the probability of the class labels. This approach performed significant results. Das et al. [21] presented the classification method for diabetes mellitus disease through data mining techniques by utilizing the J48 and Naïve Bayes algorithm on Pima Indian diabetes dataset. Poostchi et al. [22] reviewed current advancement in automated diagnosis of malaria by image processing methods and machine learning techniques. This review states that automated microscopy is a very reliable approach for malaria diagnosis. Devi et al. [23] refined a computer-aided diagnosis of malaria and erythrocytes classification through hybrid classifiers. The achievements of hybrid supervised algorithms such as *k*-NN, SVM and Naïve Bayesian techniques are performed better classifier accuracy of 98.5%.

Sahoo et al. [24] discussed a recent research application that uses the huge size of medical data while uniting multimodal data from different sources that reduce the cost and time in healthcare. Pandit and Anand [25] discussed discrete wavelet coefficients and dynamic time warping methods for diagnosis malaria from low-cost imaging devices. Discrete wavelet coefficients are used for feature vector generation and dynamic time warping method, which is used to measure the similarities and dissimilarities in red blood cells. Finally, these experimental results are shown with a classification accuracy of 91.6%. In view of the above limitations, this study target on development of computer vision and machine learning methods for the parasite detection and classification of infected and non-infected malaria regions in the microscopic images, which provides significant outcomes than the present approaches. Figure 1 showed the suggested approach which is based on computer

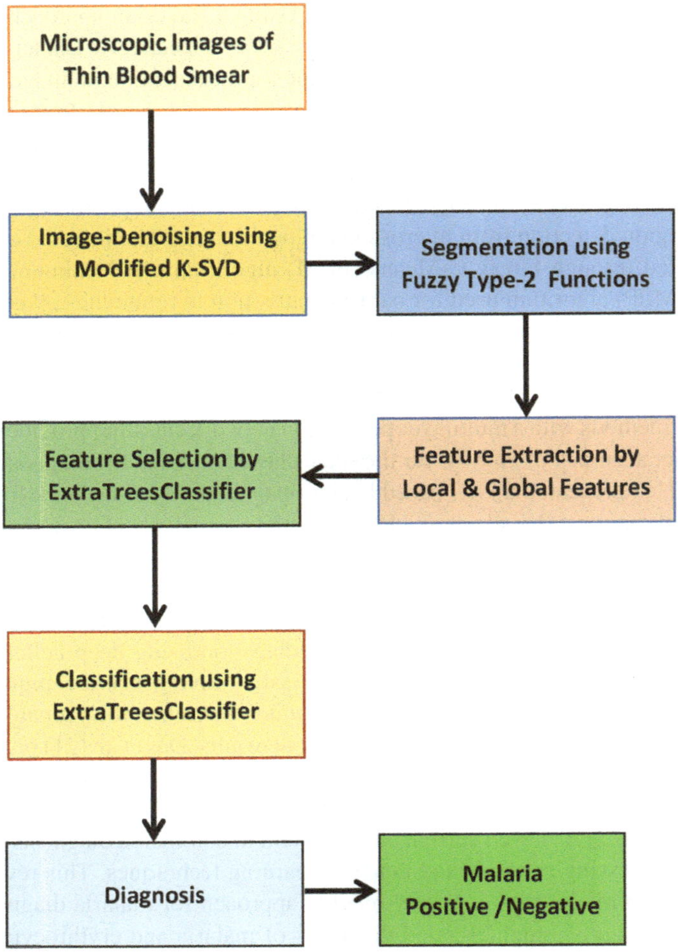

Fig. 1 Workflow diagram of the proposed work

vision and machine learning methods for the detection of malaria positive or negative cases.

3 Computer Vision and Machine Learning Methods

3.1 Dataset Collection

The database consists of a total of 400 among which 320 microscopic blood images are not publicly available and 80 are non-infected images collected from malaria repository in the U.S. National Library of Medicine [26]. These 320 images were collected from Dr. Sree Mukesh, M.D. Physician, Vivekananda Hospital, Begumpet Hyderabad, India. The blood slides were prepared under the supervision of pathologist and Dr. Sree Mukesh, M.D. Physician, Vivekananda Hospital, Begumpet Hyderabad, India. Further, these thin blood images are collected from different patients of Telangana regions with infection of Plasmodium falciparum and Plasmodium vivax.

3.2 Image Denoising

Noise occurs due to various reasons during the preparation of blood films under light microscopy. Due to this reason, several approaches have been developed and implemented on blood smears in the diagnosis of malaria. This study considers a modified *k*-SVD algorithm by adaptation of overcomplete dictionary learning [27] proposed by the same author for noise elimination in microscopic blood images using denoising phenomena. This technique performs with less computational speed and lower complexity. The denoised results are shown in Fig. 2a–d.

3.3 Cell Segmentation

Microscopic blood image consists of four different types of cells such as Leukocytes, Erythrocytes Platelets, and Plasmas. In addition, segmentation of individual blood cells for identification of parasite and its shape. In this research, Segmentation method by utilizing Einstein t-conorm and a novel fuzzy type membership function [28] utilized to the articulation of the infected blood cell region from microscopic images which is used to separate the infected parasites that gives accurate detection of cell shape and stages of malaria and it has been used for malaria diagnosis. This procedure achieved the DICE coefficient measure of 0.99%, SSIM similarity measure of 0.99% and NRMSE measure of 0.10%. Which concluded that the suggested segmentation method is a robust segmentation method. The developed segmentation algorithm is tested on various malaria test image datasets and segmented outcomes are shown in Fig. 3a–c. This technique has shown more robust segmentation approach with other techniques.

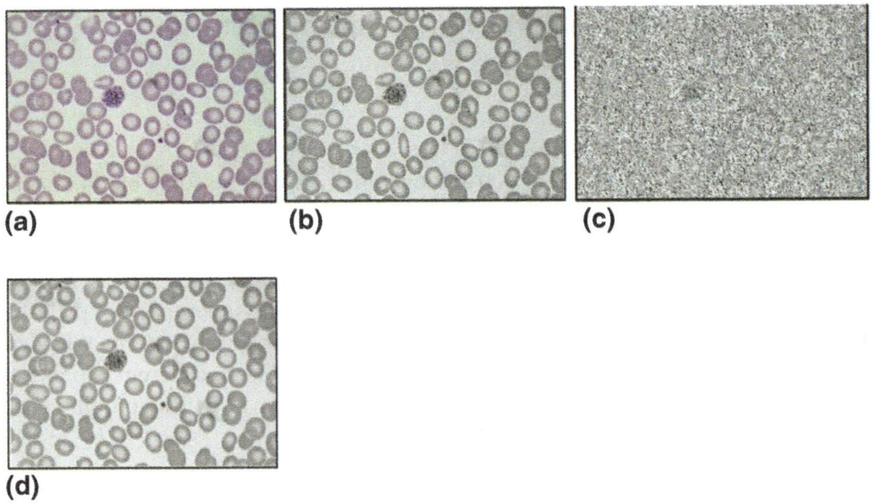

Fig. 2 (**a**) Plasmodium vivax schizont image, (**b**) grayscale image, (**c**) noise image, (**d**) denoised Image (improved *k*-SVD [27])

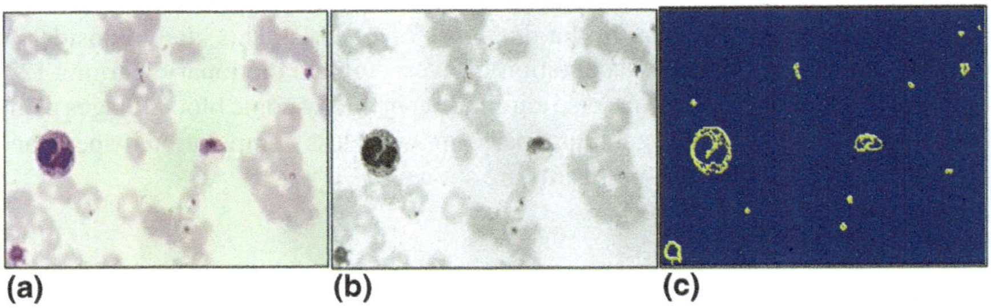

Fig. 3 (**a**) Plasmodium falciparum gametocyte image, (**b**) grayscale image, (**c**) segmented image [25]

3.4 Blood Image Feature Extraction

This research aims to extract all features of infected regions and non-infected regions of malaria parasite cells in microscopic blood images. Usually, morphological features and intensity variances are computed during the parasite detection process of the infected cell and non-infected cell of malarial images from microscopic blood images. Still, some Plasmodium falciparum species of the microscopic image don't make sense with a variety of morphological features, then this develops incompetent so that the textural features and intensity features become more stable to classify the infected regions of malaria parasites [29]. In view of these limitations, our research focuses on the detection of different life cycles of malaria parasites that are infected cells, and non-infected cells of malaria. This research conducted on a total of 603

Table 1 List of local and global feature sets are used in this experiment

S. No.	No. of features	Type of the feature set
1	f1	Shannon Entropy (01)
2	f2–f71	SURF (SpeededUp Robust Features) (70)
3	f72–f78	HuMoments (07)
4	f79–f92	Haralick Textural Features (14)
5	f93–f603	Histogram Features (512)

features by combining local and global features such as Shannon entropy, SURF, Hu-moments, Haralick, and Histograms is presented in Table 1. The features like as Shannon entropy [30], SpeedUp Robust Features (SURF) [31, 32], Hu-moments [33, 34], Haralick textural features [35, 36], and Histogram features [37]. The detailed description of the proposed feature extraction measures has been discussing in each subsection.

3.4.1 Shannon Entropy

This research, Shannon Entropy is used as a measure for feature extraction that is to be used to measure the randomness of grayscale image [30]. Also, this entropy can be used to detect sophisticated variations in the local gray level distribution values. Let us assume that $f(x, y)$ the microscopic blood cell which is malaria-infected and a non-infected image having N_i where $i = 0$ to $L - 1$ specific grayscale values. In addition, the normalized histogram computed for the region of interest (ROI) of the size of the given image ($M \times N$) and it is described as follows:

$$P_i = \frac{N_i}{M \times N} \quad (1)$$

The Shannon Entropy is defined as $S_E = -\sum_{i=0}^{L-1} P_i * \log(P_i)$ (2)

3.4.2 Speeded Up Robust Features (SURF)

Bay et al. [31, 32] given SURF is feature descriptors which compute local regions of the given image that is based on multiscale space theory and feature detectors. SURF basic concept, which is used to generate scale invariants by using local feature descriptors on a given image dataset that is employed by any feature extraction. SURF algorithm is like a scale-invariant feature transform (SIFT), it contains two major parts within the first part to find the points of interest of the image using squared cut filters and Hessian matrix. Secondly, generate feature descriptors by using localized features from these features. Usually, these are calculated by using localized feature squared image and its region about a point of interest and thus combining Haar wavelets and its return at distinct interval-based sample points. In addition, this research uses Python mahotas surf library for feature extraction which quantifies local regions of microscopic blood images. The point of interest is calculated on the whole image and these image regions are close to those interest points which are considered for further analysis.

Integral Image Detection

As a local feature detection of the microscopic image, let us consider a binary image $I(x, y)$ with the intensity at the location (x, y). To compute the integration of this image $I(x, y)$ and the integral image $I_{\sum}(x, y)$ are to be obtained and it is denoted as the sum of the input image.

$$S(x, y) = \sum_{i=0}^{n} \sum_{j=0}^{m} I(x, y) \tag{3}$$

This equation is used to compute the rectangle's four corners of the input image. In addition, speed up the counting of the second-order derivatives of Gaussian and Haar wavelet regions about the points of interest which are utilized for local feature extraction.

Interest Points—Hessian Matrix

In general, the SURF technique uses a blob detector to find the points of interest-based on the Hessian matrix (shown in Eq. 4). The determinant of this matrix is used to local variation measure about the points and these points are identified when the determination value is maximum.

$$H(f(x, y)) = \begin{bmatrix} \frac{\partial^2 f}{\partial x^2} & \frac{\partial^2 f}{\partial x \partial y} \\ \frac{\partial^2 f}{\partial x \partial y} & \frac{\partial^2 f}{\partial y^2} \end{bmatrix} \tag{4}$$

Let a point $p = (x, y)$ in a microscopic blood image $I(x, y)$ then the Hessian matrix $H(p, \sigma)$ is defined at the point p with scalar σ as follows:

$$H(p, \sigma) = \begin{pmatrix} L_{xx}(p, \sigma) & L_{xy}(p, \sigma) \\ L_{yx}(p, \sigma) & L_{yy}(p, \sigma) \end{pmatrix} \tag{5}$$

In Eq. (5) $H(p, \sigma)$ denotes the Hessian matrix and

$$L_{xx}(p, \sigma), L_{xy}(p, \sigma), L_{yx}(p, \sigma) \text{ and } L_{yy}(p, \sigma)$$

represents the convolution of Gaussian second-order derivatives w.r.t of the image.

3.4.3 Hu Moments

Hu Moments [33, 34] the intent of seven different moments of the image which are constant to image transformation. The first six moments are to be considered as invariant to scale, translation, reflection, and rotation. In addition, the seventh moment's sign changes for image reflection. In the specification of central moments and other seven Hu moments are representing as follows [33]:

$$Hm_1 = (\eta_{20} + \eta_{02}) \tag{6}$$

$$Hm_2 = (\eta_{20} - \eta_{02})^2 + 4 * \eta_{11}^2 \tag{7}$$

$$Hm_3 = (\eta_{30} - 3 * \eta_{12})^2 + ((3 * \eta_{21}) - \eta_{03})^2 \tag{8}$$

$$Hm_4 = (\eta_{30} + \eta_{12})^2 + (\eta_{21} + \eta_{03})^2 \tag{9}$$

$$\begin{aligned} Hm_5 = {} & (\eta_{30} - 3 * \eta_{12})(\eta_{30} + \eta_{12})\left[(\eta_{30} + \eta_{12})^2 - 3 * (\eta_{21} + \eta_{03})^2\right] \\ & + (3 * \eta_{21} - \eta_{03})(\eta_{21} + \eta_{03})\left[3 * (\eta_{30} + \eta_{12})^2 - (\eta_{21} + \eta_{03})^2\right] \end{aligned} \tag{10}$$

$$Hm_6 = (\eta_{20} - \eta_{02})[(\eta_{30} + \eta_{12})^2 - (\eta_{21} + \eta_{03})^2] + 4 * \eta_{11}(\eta_{30} + \eta_{12})X(\eta_{21} + \eta_{03}) \tag{11}$$

$$\begin{aligned} Hm_7 = {} & (3 * \eta_{21} - \eta_{03})(\eta_{30} + \eta_{12})\left[(\eta_{30} + \eta_{12})^2 - 3 * (\eta_{21} + \eta_{03})^2\right] - 3 * (\eta_{21} + \eta_{03})^2\Big] \\ & - (\eta_{30} - 3 * \eta_{12})(\eta_{21} + \eta_{03})\left[3 * (\eta_{30} + \eta_{12})^2 - (\eta_{21} + \eta_{03})^2\right] \end{aligned} \tag{12}$$

In Eq. (6) to Eq. (12) η_{lm} is a moment of inertia around the centroid, it is a normalized moment of order $(l + m)$ and

$$\eta_{lm} = \frac{\mu_{lm}}{\mu_{00}^{\delta}}, \text{ where } \delta = \frac{l + m}{2} + 1 \tag{13}$$

However, these Hu Moments are generating a very high order that will represent important features in our input data and it quantifies the shape of the malaria parasite.

3.4.4 Haralick Textural Features

The textural feature is one of the significant features in an object identification or region of interest in an image for image classification [35, 36], which is either microscopic image, satellite image, digital images, as fluorescence microscopy images. Haralick [36] proposed fourteen statistical features which include correlation, entropy, energy, inverse difference moments, inertia, sum averages, sum entropies, sum variances, difference averages, difference entropy's, difference variances, and other two correlation measures. In this study, Haralick texture features were used to quantify the texture features of the microscopic images for diagnosis malaria.

3.4.5 Histogram Equalization

In general, histograms relate to the intensity values of pixels in an image that is an eight-bit grayscale image which consists of 256 different bins within the range 0–255. To extract histogram features from the microscopic images used histogram processing and normalized this histogram by isolating each of its pixel values from a total number of pixel values in microscopic images. This normalized histogram can easily interpret probability functions which denotes the probability of occurrence of a grayscale intensity pixel values in an image. This approach usually increases the global feature contrast of an image, when the practical data of the image is denoted via close contrast values. From this adjustment, the intensity pixel values can be better distributed on the histogram [37].

4 Feature Selection Through ExtraTreesClassifier

After feature extraction, the feature set may have few irrelevant features that lead to overfitting in the diagnostic model. In addition, using all features of blood image may negatively influence the performance of the model which may lead the bias results of the diagnostic model. To address these issues, this study uses both textural features and morphological features which are generated in thin blood smears of microscopic images.

4.1 Feature Selection

This research used a total of 603 textural and morphological features which include local and global features, and these are generated from malaria-infected and non-infected parasite cells. ExtraTreesClassifier [38] is applied to obtain the highly influenced features using optimal strategies of ranks using the importance of weights that will improve the performance in the presence of noisy features. In addition, a meta transformer is used to select important features by using the importance of weights. This experimental study is a dataset that consists of malaria parasite-infected cells from microscopic blood images, and it contains 400 thin blood smears, which were obtained from different pathological clinics in Hyderabad, India. A total number of

Table 2 Description of the original dataset and its feature selection using ExtraTreesClassifier

S. No.	Dataset	Original size	After feature selection
1	Training set	[360, 603]	[360, 217]
2	Testing set	[40, 603]	[40, 217]

320 of 400 smears and malaria-infected images and 80 non-infected images are used for the experimental study.

The generated scheme presented in Table 2 and Figs. 4 and 5 shows the results of the top 30 important features that are extracted from thin blood smears of infected and noninfected malaria images and highly influenced features based on their ranks of weights form microscopic images which are computed through ExtraTreesClassifier feature selection in a package of Python [38].

Table 3, presents the top-20 important features which are through local and global features and each feature is computing by the Gini Index:

$$\text{Gini\_Index (Data)} = 1 - \sum_{i=1}^{n} P_i^2 \tag{14}$$

Each feature is ordered in descending order based on the Gini ranking and users can select the top k features according to his/her requirement.

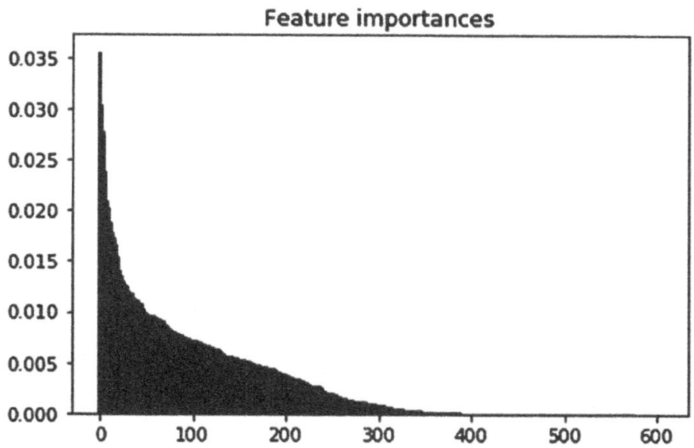

Fig. 4 Before, feature selection and its importance using ExtraTreesClassifier

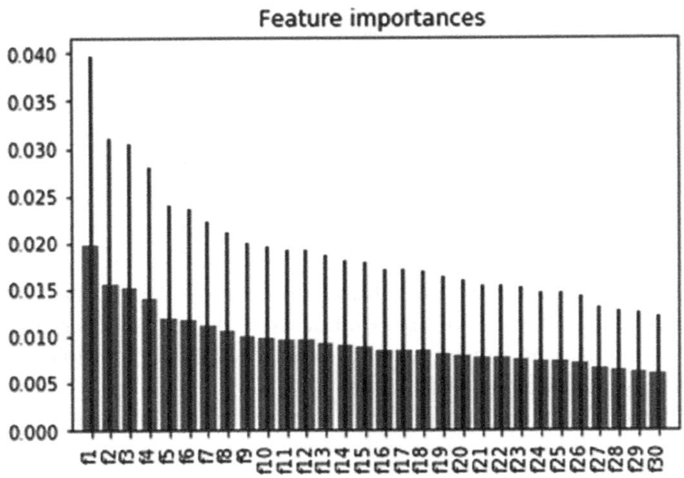

Fig. 5 After, feature selection and its feature importance using ExtraTreesClassifier

Table 3 Top-20 features are generated through ExtraTreesClassifier

S. No.	Important features	Ranking (Gini Index)	S. No.	Important features	Ranking (Gini Index)
1	Feature: 587	0.01577	11	Feature: 46	0.00977
2	Feature: 50	0.01216	12	Feature: 66	0.00943
3	Feature: 80	0.01129	13	Feature: 585	0.00883
4	Feature: 141	0.01054	14	Feature: 77	0.00876
5	Feature: 594	0.01052	15	Feature: 239	0.00842
6	Feature: 591	0.01037	16	Feature: 3	0.00824
7	Feature: 30	0.01006	17	Feature: 93	0.00778
8	Feature: 592	0.01004	18	Feature: 34	0.00756
9	Feature: 119	0.00995	19	Feature: 383	0.00735
10	Feature: 593	0.00994	20	Feature: 215	0.00730

5 Malaria Life Stages Classification Using Machine Learning Approaches

This section presents an overview of few machine learning algorithms which have been playing a vital role in the digital pathological laboratories to fascinate the initial need of clinicians for early and accurate diagnosis for malaria parasite detection and its life stages classification. In literature various classification procedures [9, 11, 12] to distinguish among parasites and other stained components of blood cells.

5.1 Extremely Randomized Trees

Geurts et al. [39] proposed another class of ensemble algorithm called Extremely Randomized Trees (ERT) for classification and regression problems. ERT classifiers are used to categorize each cell into infected versus non-infected from malaria images and enhance classifiers' performance. ERT is the flavor of Random Forests (RF) decision trees and that is an ensemble algorithm that uses a combination of decision trees without bagging by using random splits to generate different trees [36]. In ERT and RF, each tree includes a set of training datasets and a predictor, the isolate criteria start with a root node and it will process every node in the tree and tree expand until a specified depth path of the tree [39]. This implementation through meta estimator fits the samples of randomized decision trees with several sub-samples of the dataset and it controls are over-fitting.

In addition, the randomization phase is combined for choosing the key-points by the randomized choice of attributes such as in Random Forest. Choosing each node of tree at random subset of 'k' features which is to be set on the split of a node in the tree [40] and complete description of Extremely Randomized Trees as shown in Fig. 6.

Each tree is computed by no. of randomized trees for the given data point x and data D and the feature vector is generated through $f_v(x, D_{tr})$. To classify the

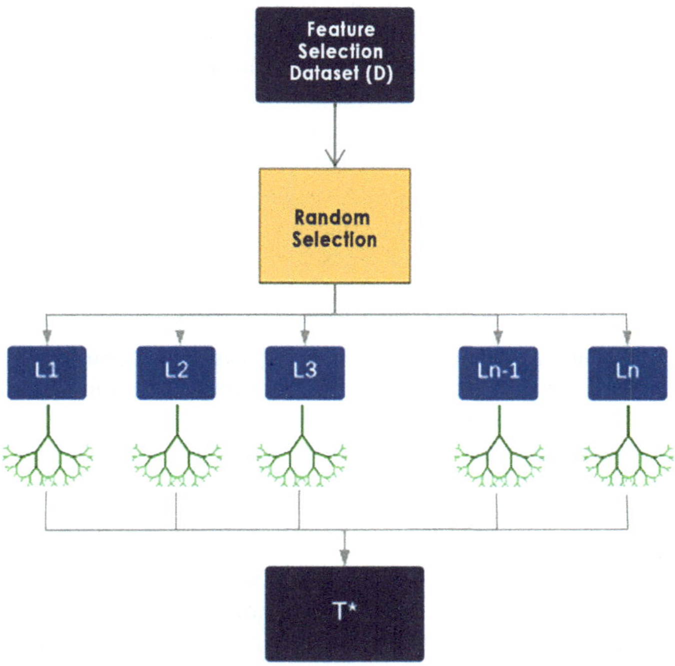

Fig. 6 Visual presentation of extremely randomized trees

multiple of class c for an n-dimensional dataset and learn a weak predictor of the tree by $p_t(c|f_v(x, D_{tr}))$. For the testing data, for an unseen data point x' is calculated by an average algorithm is producing models build on aggregating T^* different models which are obtained by a randomized algorithm and the average of probabilities of all the trees in the model as represented as follows [41]:

$$p\left(c|f_v(x', D) = \frac{1}{T^*}\sum_{t=1}^{T^*} p_t c|f_v(x', D)\right). \tag{15}$$

In our method, the parameters are selected which are 20 Extra trees, depth of the tree is generated to be 15 and min number of examples for splitting criteria chosen to be 2 in the classification task. After classification, each feature is classified into infected and uninfected of malaria from thin blood smears of microscopic images.

6 Experimentations and Results

In this section, the proposed diagnostic model and experiment analysis assessed through the database of infected malaria image datasets which were gathered from Vivekananda Hospital, Begumpet, Hyderabad, India. Noninfected malaria images were collected from the U.S. National Library of Medicine [https://ceb.nlm.nih.gov/repositories/malariadatasets/]. Achievement of the suggested model of feature extraction and selection-cum classification has been obtained to build the machine learning model for malaria parasite detection with the highest accuracy in Extremely Randomized Trees. In this study, k-fold stratified cross-validation test [42] used to overcoming the overfitting problem and generated confusion matrix on malaria life cycle based on Extremely Randomized Trees are shown in Fig. 7 and confusion matrix that help to notify the achievement of a classification model on test data of malaria life cycles such as Gametocytes, Ring forms, Schizont, Trophozoites and Non-malaria blood images for that the true values are identified for prediction. Figure 5 showed important features out of 603 are which are statistically significant measured with the Gini index, which is estimated by the Extra Trees Classifier using a package in Python [38]. Figure 7, provides the means to group the classification results into a single matrix and each row represents the predicted class of malaria stages and the column presents the actual class of malaria stages. The true predicted values are presented in a diagonal line from top-left to bottom-right of the confusion matrix and 'zero' and 'one' represent incorrectly classified as Gametocytes, Ringforms, Schizont, Trophozoites, and Non-malaria.

In this study, 400 blood images are used for the testing among 360 for training and 40 for testing. For the results of the ERT, the classifier gives an accuracy of 98.02%, precision of 98%, recall-98% and f1-score 97% for the training set. The results of Logistic regression (LR) gives an accuracy of 87.5%, the precision of 88%, recall-88% and $f1$-score 87% for the given input dataset. Table 4 showed the compared

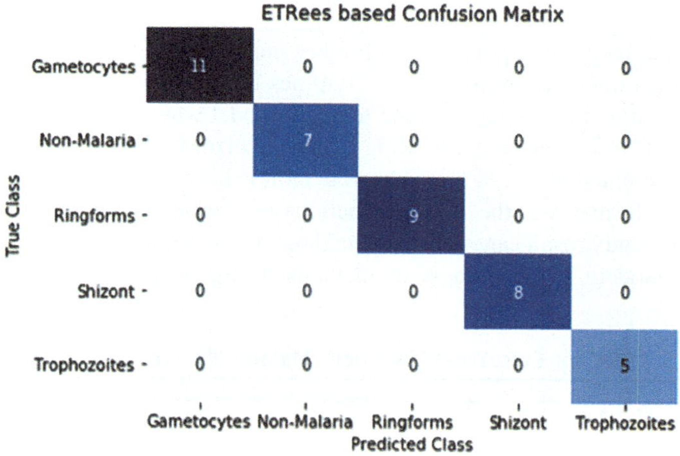

Fig. 7 ExtraTreesclassifier-based confusion matrix and its true class (y-axis) and predicted classes (x-axis)

Table 4 Performance of ExtraTreesclassifier with other classifiers on malaria life cycles and noninfected malaria

S. No	Name of the algorithm	Accuracy (%)	Precision	Recall	f1-Score
1	Proposed	98.02	0.98	0.98	0.97
2	Logistic Regression (LR)	87.5	0.88	0.88	0.87
3	Linear Discriminant Analysis (LDA)	97.5	0.98	0.97	0.97
4	CART	87.4	0.89	0.88	0.88
5	C4.5	97.5	0.98	0.97	0.97
6	k-NN (k = 5)	82.5	0.84	0.82	0.83
7	Naive Bayesian	80.00	0.80	0.80	0.79

results of the training set and testing dataset using the ERT classifier to separate malaria and non-malaria with various parameters.

The performance of Linear Discriminant Analysis (LDA) presents the accuracy of 97.5%, the precision of 98%, recall-97% and f1-score 97%. In addition, Classification and Regression Trees (CART) classifier performance provide accuracy of 87.5%, precision of 89%, recall-88% and f1-score 88% for training set. In addition, the Decision tree (C4.5) confer the same results of LDA that is the accuracy of 97.5%, the precision of 98%, recall-97% and f1-score 97%. The K-NN classification performance results in the present accuracy of 82.5%, the precision of 84%, recall-82% and f1-score 83% for the training dataset. However, for Naïve Bayesian classification performance analysis provide an accuracy of 80.00%, the precision of 80%, recall-80% and f1-score 79% for the training dataset.

The test accuracies of Extra Trees classifier are given in Table 3, which says that the suggested method outperforms with other popular classification methods. The receiver operating characteristic (ROC) outlines trade-off among the true positive and false-positive rate of our machine learning model which computed on malaria life cycles such as Gametocytes, Ring forms, Schizont, Trophozoites, and Non-malaria is presented in Fig. 8. This shows an optimistic view of Extra Trees classifier algorithm performance in the class distribution of various malaria life cycles. Lastly, comparative study results are presented in Table 5. To examine the proposed model with other modern malaria diagnosis methods. In the proposed model significant

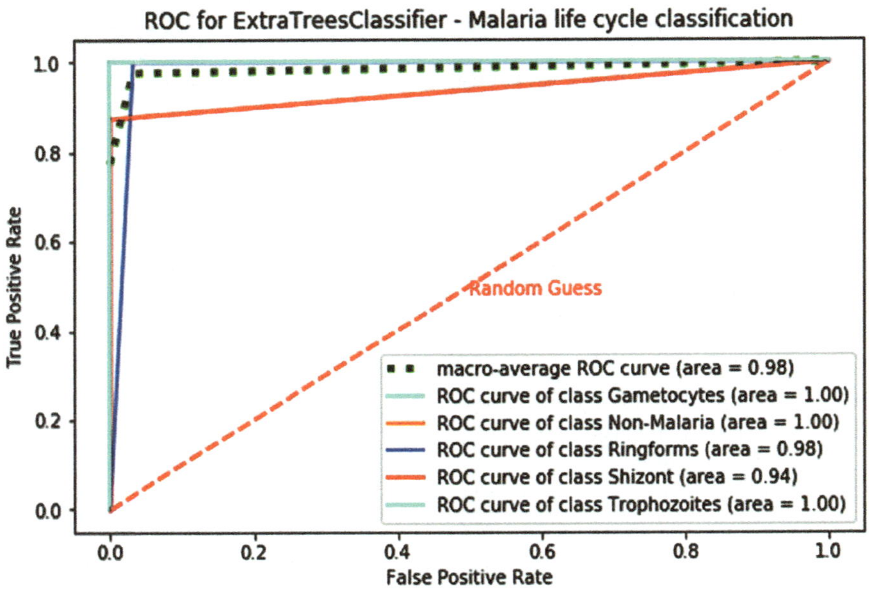

Fig. 8 ROC curves class of Malaria life cycles

Table 5 Comparative study results of our methodology with other malaria diagnosis methods

S. No.	Authors names	Type of Malaria	Sensitivity (%)	Specificity (%)
1	Diaz et al. [43]	Plasmodium Falciparum	94	99
2	Sio et al. [44]	Plasmodium Falciparum	92.50	85.23
3	Tek et al. [8]	Plasmodium Falciparum	74.58	97.5
4	Ross et al. [45]	Plasmodium Falciparum	98	85
5	Das et al. [6]	Plasmodium Falciparum, vivax	97.25	88.80
6	Proposed method	Plasmodium Falciparum, vivax	98.02	97.51

outcomes in terms of sensitivity, specificity may be used in the diagnosis of malaria for an accurate diagnosis.

7 Conclusions

This research demonstrates that computer vision and machine learning-based approaches for the diagnosis of malaria through microscopic images. However, computer vision and machine learning play an important role in digital pathology for the recognition of tissue and infected cells. In view of this, one have tried to expand a diagnostic model for malaria parasites detection along with various stages of the life cycle of malaria. The proposed computer vision and machine learning approach can quantitatively characterize both Plasmodium falciparum and Plasmodium vivax for better decision making. The computer vision approach used for denoising, segmentation, feature extraction by using local and global features, feature selection with Extra Trees Classifier and finally classification is applied to identify the different cycles of malaria that have been investigated. The proposed classification strategy can be achieved by unifying Extremely Randomized Trees. From experimentation, it is conclude that the Extremely Randomized Tree classifier performs better classification accuracy for malaria parasite detection with its cycle's classification. However, the proposed approach can characterize malaria parasite for better understanding and quality decision making. In addition, the proposed method can classify the malaria parasites into Gametocytes, Ringforms, Schizont, and Trophozoites. Furthermore, this approach is useful for quick and robust diagnosis in remote areas where medical facilities are not often available.

Limitations and Future Works: The limitation of this research is not capable to handle overlapping features of malaria stages. In the future, Deep Learning techniques would be extended to overcome overlapping problems with other stages of malaria parasites.

Acknowledgements This work was supported by the DRDO-DRL, Tezpur, Assam, India (Task No. DRLT-P1-2015/Task-64).

References

1. World Health Organization, World malaria report 2018, Nov 2018. https://www.who.int/malaria/publications/world-malaria-report-2018/report/en/. ISBN: 978 92 4
2. Greer JP, Foerster J, Rodgers GM, Paraskevas F, Glader B et al (2009) Wintrobe's clinical hematology, 12th edn. Lippincott Williams & Wilkins, Philadelphia
3. Sharma VP (1996) Reemergence of malaria in India. Indian J Med Res 103: 26–45

4. Dhiman Sunil, Veer Vijay, Dev Vas (2018) Declining transmission of Malaria in India: accelerating towards elimination, towards Malaria elimination—a leap forward, Sylvie Manguin and Vas Dev. IntechOpen. https://doi.org/10.5772/intechopen.77046
5. Devi SS, Roy A, Singha J, Sheikh SA, Laskar RH (2018) Malaria infected erythrocyte classification based on a hybrid classifier using microscopic images of thin blood smear. Multimed Tools Appl 77(1):631–660
6. Das DK, Ghosh M, Pal M, Maiti AK, Chakraborty C (2013) Machine learning approach for automated screening of malaria parasite using light microscopic images. Micron 45:97–106
7. Dhiman S, Baruah I, Singh L (2010) Military Malaria in Northeast Region of India (Review Paper). Def Sci J 60(2):213–218. https://doi.org/10.14429/dsj.60.342
8. Nicholas RE, Charles JP, David MR, Adriano GD (2006) Automated image processing method for the diagnosis and classification of malaria on thin blood smears. Med Biol Eng Comput 44(5):427–436
9. Tek FB, Dempster AG, Kale I (2006) Malaria parasite detection in peripheral blood images. In: Proceeding of the british machine vision conference, UK, pp 347–356
10. Boyd MF, Christophers R, Coggeshall LT (1949) Laboratory diagnosis of malaria infections. In: Boyd MF (ed) Malariology, vol 1. Saunders, Philadelphia, pp 177–178
11. Tek FB, Dempster AG, Kale I (2009) Computer vision for microscopy diagnosis of malaria. Malaria J 8(1):153
12. Tek FB, Dempster AG, Kale I (2010) Parasite detection and identification for automated thin blood film malaria diagnosis. Comput Vis Image Underst 114(1):21–32
13. Kumarasamy SK, Ong SH, Tan KS (2011) Robust contour reconstruction of red blood cells and parasites in the automated identification of the stages of malarial infection. Mach Vis Appl 22(3):461–469
14. Bhowmick S, Das DK, Maiti AK, Chakraborty C (2012) Computer-aided diagnosis of thalassemia using scanning electron microscopic images of peripheral blood: a morphological approach. J Med Imaging Health Inform 2(3):215–221
15. Linder N et al (2014) A malaria diagnostic tool based on computer vision screening and visualization of Plasmodium falciparum candidate areas in digitized blood smears. PLoS ONE 9(8):e104855
16. Somasekar J, Reddy BE (2015) Segmentation of erythrocytes infected with malaria parasites for the diagnosis using microscopy imaging. Comput Electr Eng 45:336–351
17. Das DK, Maiti AK, Chakraborty C (2015) Automated system for characterization and classification of malaria-infected stages using light microscopic images of thin blood smears. J Microsc 257(3):238–252
18. Das H, Jena AK, Nayak J, Naik B, Behera HS (2015) A novel PSO based backpropagation learning-MLP (PSO-BP-MLP) for classification. In: Computational intelligence in data mining, vol 2. Springer, New Delhi, pp 461–471
19. Rosado L, Correia da Costa JM, Elias D, Cardoso SJ (2016) A review of automatic malaria parasites detection and segmentation in microscopic images. Anti-Infect Agents 14(1):11–22
20. Bibin D, Nair MS, Punitha P (2017) Malaria parasite detection from peripheral blood smear images using deep belief networks. IEEE Access 5:9099–9108
21. Das H, Naik B, Behera HS (2018) Classification of diabetes mellitus disease (DMD): a data mining (DM) approach. In: Progress in computing, analytics and networking. Springer, Singapore, pp 539–549
22. Poostchi M, Silamut K, Maude RJ, Jaeger S, Thoma G (2018) Image analysis and machine learning for detecting malaria. Transl Res 194:36–55
23. Devi SS, Roy A, Singha J et al (2018) Malaria infected erythrocyte classification based on a hybrid classifier using microscopic images of a thin blood smear. Multimed Tools Appl 77:631. https://doi.org/10.1007/s11042-016-4264-7
24. Sahoo AK, Mallik S, Pradhan C, Mishra BSP, Barik RK, Das H (2019) Intelligence-based health recommendation system using big data analytics. In: Big data analytics for intelligent healthcare management, pp 227–246

25. Pandit P, Anand A (2019) Diagnosis of Malaria using wavelet coefficients and dynamic time warping. Int J Appl Comput Math 5(2):26
26. Rajaraman S, Antani SK, Poostchi M, Silamut K, Hossain MA, Maude, RJ, Jaeger S, Thoma GR (2018) Pre-trained convolutional neural networks as feature extractors toward improved Malaria parasite detection in thin blood smear images. PeerJ6:e4568. https://doi.org/10.7717/peerj.4568
27. Golla M, Rudra S (2019) A novel approach of *k*-SVD based algorithm for image denoising. In: Histopathological image analysis in medical decision making. IGI Global, pp 154–180
28. Golla M (2018) Gaussian membership function and type II fuzzy sets-based approach for edge enhancement of malaria parasites in microscopic blood images. In: International conference on ISMAC in computational vision and bio-engineering. Springer, Cham, pp 651–664
29. Savkare SS, Narote SP (2011) Automatic detection of malaria parasites for estimating parasitemia. Int J Comput Sci Secur (IJCSS) 5(3):310
30. Pharwaha APS, Sing B (2009) Shannon and non-Shannon measures of entropy for statistical texture feature extraction in digitized mammograms. In: Proceedings of WCECS, vol. I/II. San Francisco, USA, pp 1286–1291
31. Bay H, Tuytelaars T, Van Gool L (2006) Surf: speeded up robust features. In: European conference on computer vision. Springer, Berlin, Heidelberg, pp 404–417
32. Bay H, Ess A, Tuytelaars T, Van Gool L (2008) Speeded up robust features (SURF). Comput Vis Image Underst 110(3):346359
33. Hu MingKuei (1962) Visual pattern recognition by moment invariants. IRE Trans Inf Theory 8(2):179187
34. http://docs.opencv.org/modules/imgproc/doc/structural_analysis_and_shape_descriptors.html?highlight=cvmatchshapes#humoments
35. Haralick RM, Shanmugam K (1973) Textural features for image classification. IEEE Trans Syst Man Cybern 6:610–621
36. Haralick RM (1979) Statistical and structural approaches to texture. In: Proceedings of the IEEE, vol 67, no 5, pp 786–804
37. Gokhale A (2018). https://iitmcvg.github.io/summer_school/Session2/
38. Pedregosa F, Varoquaux G, Gramfort A, Michel V, Thirion B, Grisel O, ... & Vanderplas J (2011) Scikit-learn: machine learning in Python. J Mach Learn Res 12(Oct): 2825–2830
39. Geurts P, Ernst D, Wehenkel L (2006) Extremely randomized trees. Mach Learn 63(1):3–42
40. Geurts P et al (2011) Learning to rank with extremely randomized trees. In: JMLR: workshop and conference proceedings, vol 14, pp 49–61
41. Soltaninejad M, Yang G, Lambrou T, Allinson N, Jones TL, Barrick TR, ... & Ye X (2017) Automated brain tumour detection and segmentation using superpixel-based extremely randomized trees in FLAIR MRI. Int J CARS 12(2):183–203 https://doi.org/10.1007/s11548-016-1483-3
42. Patrick MT, Raja K, Miller K, Sotzen J, Gudjonsson JE, Elder JT, Tsoi LC (2019) Drug repurposing prediction for immune-mediated cutaneous diseases using a word-embedding–based machine learning approach. J Investig Dermatol 139(3):683–691
43. Díaz G, González FA, Romero E (2009) A semi-automatic method for quantification and classification of erythrocytes infected with malaria parasites in microscopic images. J Biomed Inform 42(2):296–307
44. Sio SWS, Sun W, Kumar S, Bin WZ, Tan SS et al (2007) Malaria count: an image analysis-based program for the accurate determination of parasitemia. J Microbiol Methods 68(1):11–18
45. Ross NE, Pritchard CJ, Rubin DM, Duse AG (2006) Automated image processing method for the diagnosis and classification of malaria on thin blood smears. Med Biol Eng Comput, 44(5):427–436

The manufacturer's authorised representative in the EU is Springer Nature Customer Service Centre GmbH, Europaplatz 3, 69115 Heidelberg, Germany. If you have any concerns regarding our products, please contact ProductSafety@springernature.com

Printed and bound by CPI Group (UK) Ltd, Croydon, CR0 4YY
30/09/2026
02225333-0004